SPRINGER
LAB MANUALS

Springer

Berlin
Heidelberg
New York
Barcelona
Hong Kong
London
Milan
Paris
Tokyo

JOACHIM KOCH   MICHAEL MAHLER (EDS.)

Peptide Arrays on Membrane Supports

Synthesis and Applications

With 31 Figures

Springer

Joachim Koch

Forschungsstelle Hantaviren
Heidelberger Akademie
der Wissenschaften

Current address:
Institut für Biochemie
Biozentrum N210/20
Marie-Curie-Straße 9
60439 Frankfurt am Main
Germany

e-mail:
joachim.koch@em.uni-frankfurt.de

Michael Mahler

Institut für Molekulare Genetik
Universität Heidelberg

Current address:
Pharmacia Deutschland GmbH
Munzinger Straße 7
79111 Freiburg
Germany

e-mail:
michael.mahler@pharmacia.com

ISBN 3-540-42532-2 Springer-Verlag Berlin Heidelberg New York

Library of Congress Cataloging-in-Publication Data
Peptide arrays on membane supports : synthesis and applications / Joachim Koch,
Michael Mahler (eds.).
p. cm. – (Springer lab manual)
Includes bibliographical references and index.
ISBN 3540425322 (alk. paper)
1. Protein microarray. I: Koch, Joachim, 1972- II. Mahler, Michael, 1974- III. Series.
QP551 .P394 2002
572'.65–dc21 2001049848

Springer-Verlag Berlin Heidelberg New York
a company of BertelsmannSpringer Science+Business Media GmbH

http://www.springer.de

Printed in Germany

Cover design: *design & production* GmbH, D-69121 Heidelberg
Typesetting: Kröner, D-69115 Heidelberg

SPIN 10795681 31/3130 YK – 5 4 3 2 1 0 – Printed on acid free paper

Preface

Since protein interactions are of immense interest not only for basic research but also for applied science purposes, many studies aim to shed more light on the character of the cross-talk of proteins, including antibody-antigen, protein-protein and protein-nucleic acid interactions. For this approach, a variety of different biochemical and immunological methods, such as Western blotting or ELISA, have been traditionally used. More recently, truncation studies and mutational analysis, as well as yeast two-hybrid, phage display and surface plasmon resonance techniques have been developed and applied to interaction studies in many disciplines including biochemistry, immunology, cell biology, developmental and molecular biology. Since all these methods have their individual limitations, a universal method is not available at present. In this book, we present a collection of straightforward methods based on the synthesis of oligopeptides on activated cellulose membranes that allow the investigation of protein interactions on a molecular level. This SPOT technology represents a powerful proteomics technique for a variety of applications, such as epitope mapping and the analysis of protein–protein and protein–nucleic acid interactions.

Epitope mapping has proven to be important for both basic research and diagnostic and therapeutic applications. Peptides comprising the epitope sequence(s) recognized by the majority of sera from patients with certain infectious or autoimmune diseases can serve as target antigens for new sensitive and specific diagnostic systems. For therapeutic approaches, knowledge of the exact binding site of antibodies on their antigen can further the development of new vaccination strategies against pathogens. Furthermore, such insights may also be used for the development of new therapeutic strategies in autoimmune diseases. Since signal transduction processes are highly complex and involve a large number of different interacting proteins, it is obvious that the determination of interaction sites could enlighten the character of such signal transduction cascades. Additionally, detailed knowledge of the precise interaction site(s) may lead to the creation of artificial ligands with therapeutic relevance in cancer and a variety of other dysfunctions.

The flood of data derived from the deciphering of the human genome sequence requires technologies to evaluate the biological functions of the discovered open reading frames (ORF) and to make this knowledge available for basic research and the development of therapeutic and diagnostic products. During the last few years, sophisticated micro-array systems for the investigation of protein–protein and protein–DNA interactions have been under development. The SPOT system, in combination with these high-throughput systems, may well represent a key technology to cope with the demand for information on the cross-talk of proteins.

In this book we provide a detailed overview of the technology, including insights into basic chemistry and established applications. All protocols have been written by experienced researchers in SPOT technology and have therefore been optimized for practical use. We would like to thank all of the authors for their contribution and we hope that their efforts will inspire the reader to find new implementations of this technique. This may, in turn, encourage more scientists to take advantage of the powerful tool of peptide arrays prepared with the SPOT method.

Heidelberg, Autumn 2001

Joachim Koch
Michael Mahler

Naturally Occuring Amino Acids

Amino acid	Three-letter code	One-letter code
Alanine	Ala	A
Arginine	Arg	R
Asparagine	Asn	N
Aspartic acid	Asp	D
Cysteine	Cys	C
Glutamine	Gln	Q
Glutamic acid	Glu	E
Glycine	Gly	G
Histidine	His	H
Isoleucine	Ile	I
Leucine	Leu	L
Lysine	Lys	K
Methionine	Met	M
Phylalanine	Phe	F
Proline	Pro	P
Serine	Ser	S
Threonine	Thr	T
Tryptophane	Trp	W
Tyrosine	Tyr	Y
Valine	Val	V

Contents

SPOT Synthesis – Scope of Applications

RONALD FRANK and JENS SCHNEIDER-MERGENER

Introduction

The currently very successful paradigm in scientific research of applying a systematic empirical search rather than an iterative rational design to solve complex scientific questions heavily relies on technologies that allow for a rapid and comprehensive screening of diverse types of molecular probes. Combinatorial chemical or biological synthesis applied to molecular biology, immunology and drug discovery was the technology that paved the way (Gallop et al. 1994). Massive miniaturization and automation are central and relevant topics in the further development of these technologies. A steady increase in the number of probes and samples that can be screened together with further reductions in the assay dimensions and costs readily allows for many new applications.

In 1988, Southern reported the synthesis of oligonucleotides and their arrangement in a micro-scale chessboard pattern on a planar glass surface, providing a tool in the identification of individual nucleic acid sequences in the context of the entire genome (Southern 1988) and initiating another technological breakthrough: micro-array technology. The impact of this

R. Frank (✉) (e-mail: frank@gbf.de, Tel.: +49-531-6181720, Fax: +49-531-6181795); AG Molecular Recognition, GBF (German Research Centre for Biotechnology), Mascheroder Weg 1, 38124 Braunschweig, Germany
J. Schneider-Mergener
Institut für Medizinische Immunologie, Medizinische Fakultät, Humboldt-Universität zu Berlin, Berlin, Germany
Current address: J. Schneider-Mergener, Jerini AG, Rudower Chaussee 4, 12489 Berlin, Germany

Springer Lab Manual
J. Koch, M. Mahler (Eds.) Peptide Arrays on Membrane Supports

technology is now being compared to that of the micro-electronics revolution. Other methods rapidly followed Southern's approach, such as photolithographic synthesis on glass (Pirrung et al. 1990) and SPOT synthesis on membrane supports (Frank and Güler 1990). Meanwhile, not only chemical synthesis has been utilized to generate compound probe arrays, but a series of instruments have also been developed that dispense minute volumes of solutions for arraying at feature densities reaching up to several thousand per cm^2 of any type of compound probe. Micro-array technology is a fast-developing field because it has timely enabled us to adequately utilize and exploit the enormous amount of information generated by genome and proteome research.

First presented in 1990 at the European Peptide Symposium in Barcelona (Frank et al. 1991), the SPOT synthesis method has opened up countless opportunities to synthesize and subsequently screen large arrays of synthetic peptides on planar cellulose supports. Discrete spots are arranged as *x*/*y*-arrays on membrane sheets, and each spot is individually addressable by manual or automated delivery of the appropriate reagent solutions. Although SPOT-synthesis is not as impressively miniaturized as, e.g. the Affymax (now Affimetrix) photolithographic technique, it fulfills similar demands, with the advantage of a relatively simple experimental procedure, inexpensive equipment requirements and highly flexible array and library formatting. The method permits rapid, highly parallel synthesis of huge numbers of peptides and peptide mixtures/pools (Kramer et al. 1993; Frank 1994), including a large variety of unnatural building blocks, as well as a growing range of other organic compounds. Further advantages are related to the easy adaptation to a wide range of assay and screening methods, such as binding, enzymatic and cellular assays, which allows in situ screening of the compound libraries due to the special properties of the membrane supports. Therefore, peptide arrays prepared by the SPOT technique have become quite popular tools for studying numerous aspects of molecular recognition. Already in 1991, a commercial kit for manual SPOT synthesis became available through Cambridge Research Biochemicals Ltd., and in 1992 a semi-automated SPOT synthesizer, the ASP222, was launched by ABIMED Analysen Technik.

This chapter intends to give the reader of this manual easy access to the range of diverse applications which have been described thus far (March 2001) for SPOT peptide arrays . As it is based on a comprehensive literature survey, we acknowledge the many contributions of our colleagues that have widened the scope of SPOT synthesis applications.

Directory for SPOT Peptide-Array Applications

The rational structure of this directory is built around the experimental principles of the assays that have been described. The literature survey will then cover many variations of one assay principle, which are classified according to the special biological questions approached. Entries in the directory utilize the italic numbers to cite the respective articles listed in the appended bibliography, together with a short description of the respective application given in parentheses. Technical aspects related to sample preparation, read-out systems, etc., are not addressed but will be found in the various chapters of the book. In the SPOT bibliography, only original, peer-reviewed articles are listed.

Special terms used

- Analyte — the target molecule in the assay system
- Mapping — screening of overlapping peptides derived from a protein sequence
- Deconvolution — screening, starting with more or less generic peptide libraries

Analysis by analyte binding to immobilized peptide arrays

Mapping and analysis of linear antibody epitopes

Polyclonal animal sera

- *1* (anti-CMV); *6* (anti-p60 of *L. monocytogenes*); *19* (anti-U1snRNP-C); *59* (anti-glucoamylase *of thermoanaerobacterium thermosaccharo-lyticum*); *62* (anti-interferon regulatory factor-1); *68* (anti-gliadin); *69, 83* (anti-cardiac troponin I); *118* (anti-Aβ; cross-reaction with apoE); *122* (anti-peptide from measles virus); *126* (anti-VP1-capsid of hamster polyomavirus); *138* (anti-HPEV1); *157* (anti-Vsps of *Mycoplasma bovis*)

Human sera

- *21, 49* (anti-pertussis toxin); *24* (anti-intestinal alkaline phosphatase; autoantigen in bacterial infections); *25* (anti-presenilin1); *41* (anti-gp120 of HIV1); *42* (anti-chain-3 of type IV collagen; autoantigen in Goodpasture's disease); *44* (anti-HMG-1/2box in juvenile rheumatoid arthritis); *58* (anti-SmD1 in lupus erythematosus); *66* (anti-desmoplakin I and II in erythema multiforme disease); *68* (anti-gliadin); *72* (anti-glutamic acid decarboxylase; autoantigen in insulin-dependent diabetes mellitus); *77* (anti-gG-2 of HSV-2); *81* (anti-PCM-1 autoantigen in scleroderma disease); *87* (anti-VP2 of parvovirus B19); *90* (anti-gG of HSV-1); *96* (anti-h-transaldolase; cross-reactivity with EBV and HSV capsid); *105* (anti-streptokinase); *120* (anti-sp100 in primary biliary cirrhosis); *129* (anti-CENP-A); *140* (anti-gliadin); 143 (anti-MSP1 of *P. falciparum*); *148* (anti-gG-1 of HSV-1); *151* (anti-PM/Scl-100; autoantigen in polymyositis scleroderma)

Monoclonal antibodies

- *3* (anti-MN-envelope of HIV1); *4* (F_{ab} anti-gp120 of HIV1); *5* (anti-coat protein BNYVV); *8* (anti-hIL-4); *10* (anti-P-protein of Morbillivirus); *12* (F_{ab} anti-gp41 of HIV1);); *13* (anti-p24 of HIV1); *28* (anti-gp41 of HIV1); *29* (anti-listeriolysin); *30* (anti-hFcRII/CD32); *37* (anti-h-thyro-globulin); *45* (anti-complement C3α); *52, 69, 83, 150* (anti-cardiac-troponin-I); *48* (anti-profilin); *53* (anti-RhopH3 of *Plasmodium falciparum*); *54* (anti-*Clostridium botulinum* type E neurotoxin); *56* (F_{ab} anti-p24 of HIV1); *57* (anti-prion PrP^{Sc}); *61* (anti-complement-receptor type 2, CD21); *64* (anti-troponin1); *67* (anti-yeast eIF4E); *71* (anti-EF-Tu); *74* (anti-AlaDH of *Mycobacterium tuberculosis*); *75* (anti-h-β(2)-adrenoreceptor, agonist); *77* (anti-gG-2 of HSV-2); *84* (anti-peptide, M2 acetylcholine receptor); *90* (anti-gG of HSV-1); *92* (anti-pneumolysin); *97* (F_{ab} anti-pre-S1 and pre-S2 of HBV); *98* (anti-hIL-10); *99* (anti-neurofilament in myasthenia gravis); *110* (scFv anti-large-subunit of RNA pol II of *Drosophila melanogaster*); *116* (anti-actin); *125* (mutant of scFv anti-p24 of HIV1); *134* (anti-P-τ); *136* (anti-Aβ); *139* (anti-peptide from Vβ6.2 T-cell-receptor, affinity tag); *142* (anti-peptide from *Kx* gene product); *155* (anti-tubulin-tyrosine-ligase); *159* (anti-p53; interaction with mdm2); *167* (F_{ab} anti-tobacco mosaic virus protein)

Phage-display-derived scFvs

- *82* (anti-complement C3a-receptor)

Mapping and analysis of assembled, conformational epitopes

Monoclonal antibodies

- *14, 98* (anti-hIL-10); *27* (anti-β-factorXII); *106* (scFv anti-IgG hinge region); *117* (anti hIL-10; further optimized for high affinity)

Deconvolution of linear antibody epitopes

Monoclonal antibodies

- *2, 9, 15* (anti-TGFα); *23* (F_{ab} anti-β-factorXII); *39* (anti-p24 of HIV1); *60* (D-peptide evolution, anti-p24 of HIV1); *131* (binding peptide transitions; anti-p24 of HIV1)

Mapping and analysis of antibody paratopes (CDR-derived peptides)

Antigen recognition

- *38* (mAb anti-thyroglobulin; mAb anti-lysozyme; mAb anti-angiotensinII); *70* (anti-lysozyme) *121* (mAb anti-CD4; inhibition of HIV1 transcription)

Anti-idiotypic recognition

- *141* (anti-h-thyroglobulin mAb; rabbit anti-idiotypic anti-serum)

Mapping and analysis of T-cell epitopes

MHC-binding

- *147* (HLA-DR; MSP1 peptides from *P. falciparum*)

T-cell stimulation

- *152* (immobilized spot-bound peptides)

Mapping and analysis of linear binding sites in protein-protein/peptide interactions

General

- *20* (S100C on annexin-I peptides); *26* (IgA on *group B Streptococci* β-antigen); *31* (Apeptides); *36, 80* (CaM on C-CAM peptides); *46, 145* (EVH1- domain on *Listeria monocytogenes* ActA peptides); *65,107* (TRAF1 to TRAF6 on CD40 peptides); *85, 108* (α-actinin on zyxin peptides); *86* (-, γ-tubulin association); *89* (SH3 domains of endophilin and amphiphysin on synaptojanin peptides); *101* (PDZ domain of syntrophin); *102* (Disabled1 phosphotyrosine binding domain on APP peptides); *109* (STAT3-SH3 domain on pTyr-peptides of gp130; released and rebound to MT plates via biotin); *111* (p24 on p24 peptides of HIV1); *130* (ActA of *L. monocytogenes* on p21-Arc peptides, vice versa); *133* (troponin C on h-cardiac troponin I); *135* (Mena on zyxin peptides); *144* (CD4 on human gp17 peptides); *146* (α_2-macroglobulin on β_2-microglobulin peptides); *149* (VASP/EVH1 on Fyb/SLAP2 peptides in TCR linking to actin cytoskeleton); *165* (hYAP-WW domain mutational analysis)

Receptors

- *14* (TNFon TNF-R peptides); *34* (hIL-6 on hIL6-R peptides, hIL6-R on hIL6 peptides); *43, 114* (h-transferrin on *Neisseria meningitidis* TbpB); *104* (mitochondrial import receptors Tom20, Tom22, Tom70, substrates); *119* (VEGF on VEGF-RII peptides); *140* (human secretory IgA on pneumococcal SpsA peptides)

Chaperones

- *33* (DnaK on σ^{32} peptides); *50* (DnaK substrates); *88* (SecB substrates); *91* (GroEL on Raf-1 catalytic domain peptides); *128* (mutant Dank substrates); *164* (DnaJ substrates)

Mapping and analysis of assembled conformational binding sites in protein-protein/peptide interactions

- *78* (interleukin-10 on interleukin-10 receptor)

Deconvolution of linear binding sites in protein-protein/peptide interactions

- *9* (TGFβ); *35* (streptavidin); *51* (A); **63** (PDZ domain of syntrophin); *153* (Factor VIII; affinity purification)

Enzyme inhibition

Proteases

- *124* (elastase; OMTKY3 peptide)

Protein kinases

- *127* (PKG, by deconvolution, membrane-permeant)

Protein/peptide-nucleic acid interactions

- *112* (oligonucleotide duplex on EcoRII peptides)

Peptide nucleic acid (PNA) nucleic acid interactions

- *47, 100* (hybridization studies)

Peptide interactions with small ligands

Mapping

- *132* (Heme-binding CDR peptides from anti-heme mAb)

Deconvolution

- *2* (Ag, Fe, Tc, Ca, Ni, Mn); *9* (Ni, ^{99m}Tc); *19* (^{99m}Tc)

Analysis by chemical/enzymatic transformation of immobilized peptide arrays

Chemical transformations

- *17* (chemical ligation); *93* (glycation, AGE cross-linking)

Enzymatic transformations

Mapping and analysis of protein kinase substrates

- *16* (*lyn* kinase); *32* (PKA, PKC; CKI, CKII); *55, 80* (PKC); *156* (DYRK1A)

Deconvolution of protein kinase substrates

- *11, 103* (PKA,PKG); *76* (enzyme 1 of bacterial phosphotransferase system); *154* (CDPK-1)

Mapping and analysis of protease substrates

- *24* (elastase; OMTKY3 peptide); *162* (protease OmpT)

Analysis of bound analyte

This applications refers to the use of peptide arrays as tools for multiple-affinity enrichment and isolation of bound analyte.

Antibodies

- *113, 115* (anti-bacteriophage ϕ29 connector protein; protein topology studies)

Proteins

- *22* (cell extract; cytoplasmic TNF-receptor interaction with intracellular proteins)

Analysis by the activity of cleaved solution-phase peptide arrays

T-cell epitopes

- *7* (DKP-release; M-protein of Influenza H7N1); *73* (DKP-release; p60 of *L. monocytogenes*); *94, 160* (ammonia release; OspA of *Borrelia burgdorferi*; cross-reactivity with self peptides); *95* (ammonia release; myelin basic protein; cross-reactivity with microbial peptides); *163* (ammonia release; ELISPOT; lysteriolysine, p60 of *L. monocytogenes*)

Enzymes

- *124* (elastase; OMTKY3 peptide; ammonia release)

Cells

- *79* (SDF-1 mapping; ammonia release; HIV1 infectivity)

De novo protein design

- *158* (heme-binding 4-helix bundles); *166* (copper-binding 4-helix bundles)

Miscellaneous

- *40* (spot-bound peptides for immunization); *161* (peptide-dye conjugates)

The "soft" literature even covers a much wider scope of applications than collected here. So, in the near future we can expect many more new entries to this directory.

Chemical synthesis and screening of non-peptidic compound arrays is not covered by this review, as it is outside the topic of this manual; but the respective applications are of increasing interest and relevance (Borman 2000). This is particularly true for the rapidly growing, new field of ligand-based discovery of gene function: "chemical genomics/proteomics" (Stockwell 2000).

Appendix: SPOT Bibliography

1. Frank R (1992) Spot-synthesis: an easy technique for the positionally addressable, parallel chemical synthesis on a membrane support. Tetrahedron 48:9217–9232

2. Kramer A, Volkmer-Engert R, Malin R, Reineke U, Schneider-Mergener J (1993) Simultaneous synthesis of peptide libraries on single resin and continuous cellulose membrane supports: examples for the identification of protein, metal and DNA binding peptide mixtures. Peptides Res 6:314–319

3. Jellis CL, Cradick TJ, Rennert P, Salinas P, Boyd J, Amirault T, Gray GS (1993) Defining critical residues in the epitope for a HIV-neutralizing monoclonal antibody using phage display and peptide array technologies. Gene 137:63–68

4. Rini JM, Stanfield RL, Stura EA, Salinas P, Profy AT, Wilson IA (1993) Crystal structure of a human immunodeficiency virus type 1 neutralizing antibody, 50.1, in complex with its V3 loop peptide antigen. Proc Natl Acad Sci USA 90:6325–6329
5. Commandeur U, Koenig R, Manteuffel R, Torrance L, Lüddecke P, Frank R (1994) Location, size and complexity of epitopes on the coat protein of beet necrotic yellow vein virus studied by means of synthetic overlapping peptides. Virology 198:282–287
6. Bubert A, Schubert P, Köhler S, Frank R, Goebel W (1994) Synthetic peptides derived from the p60 protein of *Listeria monocytogenes* as antigens for the generation of polyclonal antibodies specific for secreted cell-free *L. monocytogenes* p60 proteins. Appl Environ Microbiol 60:120- 3127
7. Adler S, Frank R, Lanzavecchia A, Weiss S (1994) T cell epitope analysis with peptides simultaneously synthesized on cellulose membranes: fine mapping of two DQ dependent epitopes. FEBS Lett 352:167–170
8. Reusch P, Arnold S, Heusser C, Wagner K, Weston B, Sebald W (1994) Neutralizing monoclonal antibodies define two different functional sites in human interleukin-4. Eur J Biochem 222:491–499
9. Kramer A, Schuster A, Reineke U, Malin R, Volkmer-Engert R, Landgraf C, Schneider-Mergener J (1994) Combinatorial cellulose-bound peptide libraries: screening tools for the identification of peptides that bind ligands with predefined specificity. Methods (A Companion to Methods in Enzymology) 6:388–395
10. Martens W, Greiser-Wilke I, Harder T, Dittmar K, Frank R, Örvell C, Moennig V, Liess B (1995) Spot synthesis of overlapping peptides on paper membrane supports enable the identification of linear monoclonal antibody binding determinants on morbillivirus phosphoproteins. Vet Microbiol 44:289–298
11. Tegge W, Frank R, Hofmann F, Dostmann RG (1995) Determination of cyclic nucleotide-dependent protein kinase substrate specificity with peptide libraries on cellulose paper. Biochemistry 34:10569–10577
12. Stigler R-D, Rüker F, Katinger D, Elliott G, Höhne W, Henklein P, X.Ho J, Keeling K, Carter DC, Nugel E, Kramer A, Porstmann T, Schneider-Mergener J (1995) Interaction between a Fab fragment against gp41 of human immunodeficiency virus 1 and its peptide epitope: characterization using a peptide epitope library and molecular modeling. Protein Eng 8:471–479
13. Volkmer-Engert R, Ehrhard B, Hellwig J, Kramer A, Höhne W, Schneider-Mergener J (1995) Preparation, analysis and antibody binding studies of simultaneously synthesized soluble and cellulose-bound HIV-1 p24 peptide epitope libraries. Lett Peptide Sci 1:243–254
14. Reinecke U, Sabat R, Kramer A, Stigler R-D, Seifert M, Michel T, Volk, H-D, Schneider-Mergener J (1995) Mapping protein-protein contact sites using cellulose-bound peptide scans. Mol Div 1:141–148
15. Kramer A, Vakalopoulou E, Schleuning W-D, Schneider-Mergener J (1995) A general route to fingerprint analyses of peptide-antibody interactions using a clustered amino acid peptide library: comparison with a phage display library. Mol Immunol 32:459–465
16. Szallasi Z, Denning MF, Chang E-Y, Rivera J, Yuspa SH, Lehel C, Ohla Z, Anderson WB, Blumberg PM (1995) Development of a rapid approach to

identification of tyrosine phosphorylation sites: application to PKCδ phosphorylated upon activation of the high affinity receptor for IgE in rat basophilic leukimia cells. Biochem Biophys Res Commun 214:888–894

17. Tam JP, Rao C, Liu C-F, Shao J (1995) Specificity and formation of unusual amino acids of an amide ligation strategy for unprotected peptides. Int J Peptide Protein Res 45:209–216
18. Malin R, Steinbrecher A, Semmler W, Noll B, Johannsen B, Frömmel C, Höhne W, Schneider-Mergener J (1995) Identification of technetium-99 binding peptides using cellulose-bound combinatorial peptide libraries. J Am Chem Soc 117:11821–11822
19. Halimi H, Dumortier H, Briand JP, Muller S (1996) Comparison of two different methods using overlapping synthetic peptides for localizing linear B cell epitopes in the U1 snRNP- C autoantigen. J Immunol Methods 199:77–85
20. Seemann J, Weber K, Gerke V (1996) Structural requirements for annexin I-S100C complex-formation. Biochem J 319:123–129
21. von Olleschik-Elbheim L, el Baya A, Schmidt MA (1996) Quantification of immunological membrane reactions employing a digital desk top scanner and standard graphics software. J Immunol Methods 197:181–186
22. Gao B, Esnouf MP (1996) Elucidation of the core residues of an epitope using membrane-based combinatorial peptide libraries. J Biol Chem 271: 24634–24638
23. Adam-Klages S, Adam D, Wiegmann K, Struve S, Kolanus W, Schneider-Mergener J, Kronke M (1996) FAN, a novel WD-repeat protein, couples the p55 TNF-receptor to neutral sphingomyelinase. Cell 86:937–947
24. Kolbus N, Beuche W, Felgenhauer K, Mader M (1996) Definition of a discontinuous immunodominant epitope of intestinal alkaline phosphatase, an autoantigen in acute bacterial infections. Clin Immunol Immunopathol 80:298–306
25. Thinakaran G, Borchelt DR, Lee MK, Slunt HH, Spitzer L, Kim G, Ratovitsky T, Davenport F, Nordstedt C, Seeger M, Hardy J, Levey AI, Gandy SE, Jenkins NA, Copeland NG, Price DL, Sisodia SS (1996) Endoproteolysis of presenilin 1 and accumulation of processed derivatives in vivo. Neuron 17:181–190
26. Jerlstrom PG, Talay SR, Valentin-Weigand P, Timmis KN, Chhatwal GS (1996) Identification of an immunoglobulin A binding motif located in the beta-antigen of the c protein complex of group B streptococci. Infect Immun 64:2787–2793
27. Gao B, Esnouf MP (1996) Multiple interactive residues of recognition – Elucidation of discontinuous epitopes with linear peptides. J Immunol 157:183–188
28. Purtscher M, Trkola A, Grassauer A, Schulz PM, Klima A, Dopper S, Gruber G, Buchacher A, Muster T, Katinger H (1996) Restricted antigenic variability of the epitope recognized by the neutralizing gp41 ant body 2F5. Aids 10:587–593
29. Darji A, Niebuhr K, Hense M, Wehland J, Chakraborty T, Weiss S (1996) Neutralizing monoclonal antibodies against listeriolysin: Mapping of epitopes involved in pore formation. Infect Immun 64:2356–2358
30. Weinrich V, Sondermann P, Bewarder N, Wissel K, Frey J (1996) Epitope mapping of new monoclonal antibodies recognizing distinct human FcRII (CD32) isoforms. Hybridoma 15:109–116

31. Tjernberg LO, Naslund J, Lindqvist F. Johansson J, Karlstrom AR, Thyberg J, Terenius L, Nordstedt C (1996) Arrest of beta-amyloid fibril formation by a pentapeptide ligand. J Biol Chem 271:8545–8548
32. Toomik R, Edlund M, Ek P, Obrink B, Engstrom L (1996) Simultaneously synthesized peptides on continuous cellulose membranes as substrates for protein kinases. Peptide Res 9:6–11
33. McCarty JS, Rüdiger S, Schönfeld HJ, Schneider-Mergener J, Nakahigashi K, Yura T, Bukau B (1996) Regulatory region C of the *E.-coli* heat shock transcription factor, sigma(32) constitutes a DnaK binding site and is conserved among eubacteria. J Mol Biol 256:829–837
34. Weiergraber O, Schneider-Mergener J, Grotzinger J, Wollmer A, Kuster A, Exner M, Heinrich PC (1996) Use of immobilized synthetic peptides for the identification of contact sites between human interleukin-6 and its receptor. FEBS Lett. 379:122–126
35. Schmidt TGM, Koepke J, Frank R, Skerra A (1996) Molecular interaction between the Strep-tag affinity peptide and its cognate target, streptavidin. J Mol Biol 255:753–766
36. Edlund M, Blikstad I, Obrink B (1996) Calmodulin binds to specific sequences in the cytoplasmic domain of C-CAM and down-regulates C-CAM self-association. J Biol Chem 271:1393–1399
37. Molina F, Pau B, Granier C (1997) The PGK epitope of human thyroglobulin: a molecular marker of alternatively spliced thyroglobulin molecules? Lett Pept Sci 4:201–205
38. Laune D, Molina F, Ferrieres G, Mani JC, Cohen P, Simon D, Bernardi T, Piechaczyk M, Pau B, Granier C (1997) Systematic exploration of the antigen binding-activity of synthetic peptides isolated from the variable regions of immunoglobulins. J Biol Chem 272:30937–30944
39. Kramer A, Keitel T, Winkler K, Stocklein W, Höhne W, Schneider-Mergener J (1997) Molecular basis for the binding promiscuity of an anti-p24 (HIV-1) monoclonal antibody. Cell 91:799–809
40. Strutzberg K, Franz B, Gerlach GF (1997) Interference of peptides and specific antibodies with the function of the *Actinobacillus pleuropneumoniae* transferrin-binding protein. Infect Immun 65:5346–5348
41. Schreiber M, Wachsmuth C, Müller H, Odemuyiwa S, Schmitz H, Meyer S, Meyer B, Schneider-Mergener J (1997) The V3-directed immune response in natural human immunodeficiency virus type 1 infection is predominantly directed against a variable, discontinuous epitope presented by the gp120 V3 domain. J Virol 71:9198–9205
42. Levy JB, Coulthart A, Pusey CD (1997) Mapping B cell epitopes in Goodpasture's disease. J Am Soc Nephrol 8:1698–1705
43. Renauld-Mongenie G, Poncet D, von Olleschik-Elbheim L, Cournez T, Mignon M, Schmidt MA, Quentin-Millet MJ (1997) Identification of human transferrin-binding sites within meningococcal transferrin-binding protein B. J Bacteriol 179:6400–6407
44. Jung F, Neuer G, Bautz FA (1997) Antibodies against a peptide sequence located in the linker region of the HMG-1/2 box domains in sera from patients with juvenile rheumatoid arthritis. Arthritis Rheum 40:1803–1809
45. Vogel U, Weinberger A, Frank R, Müller A, Kohl J, Atkinson JP, Frosch M (1997) Complement factor C3 deposition and serum resistance in

isogenic capsule and lipooligosaccharide sialic acid mutants of serogroup B *Neisseria meningitidis*. Infect Immun 65:4022–4029
46. Niebuhr K, Ebel F, Frank R, Reinhard M, Domann E, Carl UD, Walter U, Gertler FB, Wehland J, Chakraborty T (1997) Novel proline-rich motif present in ActA of Listeria monocytogenes and cytoskeletal proteins is the ligand for the EVH1 domain, a protein module present in the Ena/VASP family. EMBO J 16:5433–5444
47. Weiler J, Gausepohl H, Hauser N. Jensen ON, Hoheisel JD (1997) Hybridisation based DNA screening on peptide nucleic acid (PNA) oligomer arrays. Nucleic Acids Res 25:2792–2799
48. Mayboroda O, Schluter K. Jockusch BM (1997) Differential colocalization of profilin with microfilaments in PtK2 cells. Cell Motil Cytoskeleton 37:166–177
49. von Olleschik-Elbheim L, el Baya A, Schmidt MA (1997) Membrane anchored synthetic peptides as a tool for structure- function analysis of pertussis toxin and its target proteins. Adv Exp Med Biol 419:87–91
50. Rüdiger S, Germeroth L, Schneider-Mergener J, Bukau B (1997) Substrate specificity of the DnaK chaperone determined by screening cellulose-bound peptide libraries. EMBO J 16:1501–1507
51. Tjernberg LO, Lilliehook C, Callaway DJE, Naslund J, Hahne S, Thyberg J, Terenius L, Nordstedt C (1997) Controlling amyloid beta-peptide fibril formation with protease-stable ligands. J Biol Chem 272:12601–12605
52. Rama D, Calzolari C, Granier C, Pau B (1997) Epitope localization of monoclonal antibodies used in human troponin I immunoenzymometric assay. Hybridoma 16:153–157
53. Doury JC, Goasdoue JL, Tolou H, Martelloni M, Bonnefoy S, Mercereau-Puijalon O (1997) Characterisation of the binding sites of monoclonal antibodies reacting with the *Plasmodium falciparum* rhoptry protein RhopH3. Mol Biochem Parasitol 85:149–159
54. Kubota T, Watanabe T, Yokosawa N, Tsuzuki K, Indoh T, Moriishi K, Sanda K, Maki Y, Inoue K, Fujii N (1997) Epitope regions in the heavy chain of *Clostridium botulinum* type E neurotoxin recognized by monoclonal antibodies. Appl. Environ Microbiol 63:1214–1218
55. Toomik R, Ek P (1997) A potent and highly selective peptide substrate for protein kinase C assay. Biochem J 322:455–460
56. Keitel T, Kramer A, Wessner H, Scholz C, Schneider-Mergener J (1997) Crystallographic analysis of an anti-p24 (HIV-1) monoclonal antibody cross reactivity and polyspecificity. Cell 91:811–820
57. Korth C, Stierli B, Streit P, Moser M, Schaller O, Fischer R, Schulz-Schäffer W, Kretzschar H, Raeber A, Braun U, Ehrensperger F, Hornemann S, Glockshuber R, Riek R, Billeter M, Wüthrich K, Oesch B (1997) Prion (PrPsc)-specific epitope defined by a monoclonal antibody. Nature 390:74–77
58. Riemekasten G, Marell J, Trebeljahr G, Klein R, Hausdorf G, Haupl T, Schneider-Mergener J, Burmester GR, Hiepe F (1998) Novel epitope on the C-terminus of SmD1 is recognized by the majority of sera from patients with systemic lupus erythematosus. J Clin Invest 102:754–763
59. Ducki A, Grundmann O, Konermann L, Mayer F, Hoppert M (1998) Glucoamylase from *Thermoanaerobacterium thermosaccharolyticum*: Sequence studies and analysis of the macromolecular architecture of the enzyme. J Gen Appl Microbiol 44:327–335

60. Kramer A, Stigler RD, Knaute T, Hoffmann B, Schneider-Mergener J (1998) Stepwise transformation of a cholera toxin and a p24 (HIV-1) epitope into D-peptide analogs. Protein Eng 11:941–948
61. Prodinger WM, Schwendinger MG, Schoch J, Köchle M, Larcher C, Dierich MP (1998) Characterization of C3dg binding to a recess formed between short consensus repeats 1 and 2 of complement receptor type 2 (CR2; CD21). J Immunol 161:4604–4610
62. Schaper F, Kirchhoff S, Posern G, Koster M, Oumard A, Sharf R, Levi BZ, Hauser H (1998) Functional domains of interferon regulatory factor 1 (IRF-1). Biochem J 335:147–157
63. Schultz J, Hoffmüller U, Krause G, Ashurst J, Macias MJ, Schmieder P, Schneider-Mergener J, Oschkinat H (1998) Specific interactions between the syntrophin PDZ domain and the voltage-gated sodium channels. Nat Struct Biol 5:19–24
64. Filatov VL, Katrukha AG, Bereznikova AV, Esakova TV, Bulargina TV, Kolosova OV, Severin ES, Gusev NB (1998) Epitope mapping of anti-troponin I monoclonal antibodies. Biochem Mol Biol Int 45:1179– 1187
65. Pullen SS, Miller HG, Everdeen DS, Dang TTA, Crute JJ, Kehry MR (1998) CD40-Tumor necrosis factor receptor-associated factor (TRAF) interactions: regulation of CD40 signaling through multiple TRAF binding sites and TRAF hetero-oligomerization. Biochemistry 37:11836–11845
66. Foedinger D, Elbe-Burger A, Sterniczky B, Lackner M, Horvat R, Wolff K, Rappersberger K (1998) *Erythema multiforme* associated human autoantibodies against desmoplakin I and II: biochemical characterization and passive transfer studies into newborn mice. J Invest Dermatol 111:503–510
67. Ptushkina M, von der Haar T, Vasilescu S, Frank R, Birkenhager R, McCarthy JEG (1998) Cooperative modulation by eIF4G of eIF4E-binding to the mRNA 5' cap in yeast involves a site partially shared by p20. EMBO J 17:4798–4808
68. Osman AA, Uhlig H, Thamm B, Schneider-Mergener J, Mothes T (1998) Use of the phage display technique for detection of epitopes recognized by polyclonal rabbit gliadin antibodies. FEBS Lett 433:103–107
69. Larue C, Ferrieres G, Laprade M, Calzolari C, Granier C (1998) Antigenic definition of cardiac troponin I. Clin Chem Lab Med 36:361–365
70. Laune D, Pau B, Granier C (1998) Peptide models of immunological recognition: paratope dissection by multiple peptide synthesis. Clin Chem Lab Med 36:367–371
71. Baensch M, Frank R, Kohl J (1998) Conservation of the amino-terminal epitope of elongation factor Tu in eubacteria and archaea. Microbiology (UK) 144:2241–2246
72. Rharbaoui F, Granier C, Kellou M, Mani JC, van Endert P, Madec AM, Boitard C, Pau B, Bouanani M (1998) Peptide specificity of high-titer anti-glutamic acid decarboxylase (GAD)65 autoantibodies. Immunol Lett 62:123–130
73. Geginat G, Lalic M, Kretschmar M, Goebel W, Hof H, Palm D, Bubert A (1998) Th1 cells specific for a secreted protein of *Listeria* monocytogenes are protective in vivo. J Immunol 160:6046–6055
74. Hutter B, Singh M (1998) Host vector system for high-level expression and purification of recombinant, enzymatically active alanine dehydrogenase of *Mycobacterium tuberculosis*. Gene 212:21–29

75. Lebesgue D, Wallukat G, Mijares A, Granier C, Argibay J, Hoebeke J (1998) An agonist-like monoclonal antibody against the human beta(2)-adrenoceptor. Eur J Pharmacol 348:123–133
76. Mukhija S, Germeroth L, Schneider-Mergener J, Erni B (1998) Identification of peptides inhibiting enzyme I of the bacterial phosphotransferase system using combinatorial cellulose-bound peptide libraries. Eur J Biochem 254:433–438
77. Liljeqvist JA, Trybala E, Svennerholm B, Jeansson S, Sjogren-Jansson E, Bergstrom T (1998) Localization of type-specific epitopes of herpes simplex virus type 2 glycoprotein G recognized by human and mouse antibodies. J Gen Virol 79:1215–1224
78. Reineke U, Sabat R, Volk HD, Schneider-Mergener J (1998) Mapping of the interleukin-10/interleukin-10 receptor combining site. Protein Sci 7:951–960
79. Heveker N, Montes M, Germeroth L, Amara A, Trautmann A, Alizon M, Schneider-Mergener J (1998) Dissociation of the signalling and antiviral properties of SDF-1-derived small peptides. Curr Biol 8:369–376
80. Edlund M, Wikstrom K, Toomik R, Ek P, Obrink B (1998) Characterization of protein kinase C-mediated phosphorylation of the short cytoplasmic domain isoform of C-CAM. FEBS Lett 425:166–170
81. Bao LM, Varden CE, Zimmer WE, Balczon R (1998) Localization of autoepitopes on the PCM-1 autoantigen using scleroderma sera with autoantibodies against the centrosome. Mol Biol Rep 25:111–119
82. Hawlisch H, Frank R, Hennecke M, Baensch M, Sohns B, Arseniev L, Bautsch W, Kola A, Klos A, Kohl J (1998) Site-directed C3a receptor antibodies from phage display libraries. J Immunol 160:2947–2958
83. Ferrieres G, Calzolari C, Mani JC, Laune D, Trinquier S, Laprade M, Larue C, Pau B, Granier C (1998) Human cardiac troponin I: precise identification of antigenic epitopes and prediction of secondary structure. Clin Chem 44:487–493
84. Elies R, Fu LXM, Eftekhari P, Wallukat G, Schulze W, Granier C, Hjalmarson A, Hoebeke J (1998) Immunochemical and functional characterization of an agonist- like monoclonal antibody against the M2 acetylcholine receptor. Eur J Biochem 251:659–666
85. Drees BE, Andrews KM, Beckerle MC (1999) Molecular dissection of zyxin function reveals its involvement in cell motility. J Cell Biol 147:1549–1559
86. Llanos R, Chevrier V, Ronjat M, Meurer-Grob P, Martinez P, Frank R, Bornens M, Wade RH, Wehland J, Job D (1999) Tubulin binding sites on gamma-tubulin: identification and molecular characterization. Biochemistry 38:15712–15720
87. Kaikkonen L, Lankinen H, Harjunpaa I, Hokynar K, Soderlund-Venermo M, Oker-Blom C, Hedman L, Hedman K (1999) Acute-phase-specific heptapeptide epitope for diagnosis of parvovirus B19 infection. J Clin Microbiol 37:3952–3956
88. Knoblauch NTM, Rüdiger S, Schönfeld HJ, Driessen AJM, Schneider-Mergener J, Bukau B (1999) Substrate specificity of the SecB chaperone. J Biol Chem 274:34219–34225
89. Cestra G, Castagnoli L, Dente L, Minenkova O, Petrelli A, Migone N, Hoffmüller U, Schneider-Mergener J, Cesareni G (1999) The SH3 domains

of endophilin and amphiphysin bind to the proline-rich region of synaptojanin 1 at distinct sites that display an unconventional binding specificity. J Biol Chem 274:32001–32007

90. Rekabdar E, Tunback P, Liljeqvist JA, Bergstrom T (1999) Variability of the glycoprotein G gene in clinical isolates of herpes simplex virus type 1. Clin Diagn Lab Immunol 6:826–831

91. Ho MF, Wilson BA, Peterson JW (1999) Bacterially expressed Raf-1 catalytic domain is highly associated with GroEL. J Chin Chem Soc 46:735–742

92. Jacobs T, Cima-Cabal MD, Darji A, Mendez FJ, Vazquez F, Jacobs AAC, Shimada Y, Ohno-Iwashita Y, Weiss S, de los Toyos JR (1999) The conserved undecapeptide shared by thiol-activated cytolysins is involved in membrane binding. FEBS Lett 459:463–466

93. Munch G, Schicktanz D, Behme A, Gerlach M, Riederer P, Palm D, Schinzel R (1999) Amino acid specificity of glycation and protein-AGE crosslinking reactivities determined with a dipeptide SPOT library. Nat Biotechnol 17:1006–1010

94. Maier B, Molinger M, Cope AP, Fugger L, Schneider-Mergener J, Sonderstrup G, Kramer A, Kamradt T (1999) Multiple cross-reactive self-ligands for Borrelia burgdorferi outer surface protein A (OspA)-specific HLA-DR4-restricted T cells. Zentralbl Bakteriol Int J Med Microbiol Virol Parasitol Infect Dis 289:673–673

95. Grogan JL, Kramer A, Nogai A, Dong LY, Ohde M, Schneider-Mergener J, Kamradt T (1999) Cross-reactivity of myelin basic protein-specific T cells with multiple microbial peptides: experimental autoimmune encephalomyelitis induction in TCR transgenic mice. J Immunol 163:3764–3770

96. Esposito M, Venkatesh V, Otvos L, Weng ZP, Vajda S, Banki K, Peri A (1999) Human transaldolase and cross-reactive viral epitopes identified by autoantibodies of multiple sclerosis patients. J Immunol 163:4027–4032

97. Kuttner G, Kramer A, Schmidtke G, Giessmann E, Dong L, Roggenbuck D, Scholz C, Seifert M, Stigler RD, Schneider-Mergener J, Porstmann T, Hohne W (1999) Characterization of neutralizing anti-pre-S1 and anti-pre-S2 (HBV) monoclonal antibodies and their fragments. Mol Immunol 36:669–683

98. Reineke U, Schneider-Mergener J, Glaser RW, Stigler RD, Seifert M, Volk HD, Sabat R (1999) Evidence for conformationally different states of interleukin- 10: binding of a neutralizing antibody enhances accessibility of a hidden epitope. J Mol Recognit 12:242–248

99. Schultz A, Hoffacker V, Wilisch A, Nix W, Gold R, Schalke B, Tzartos S, Muller-Hermelink HK, Marx A (1999) Neurofilament is an autoantigenic determinant in myasthenia gravis. Ann Neurol 46:167–175

100. Matysiak S, Hauser NC, Wurtz S, Hoheisel JD (1999) Improved solid supports and spacer/linker systems for the synthesis of spatially addressable PNA-libraries. Nucleosides Nucleotides 18:1289–1291

101. Hoffmüller U, Russwurm M, Kleinjung F, Ashurst J, Oschkinat H, Volkmer-Engert R, Koesling D, Schneider-Mergener J (1999) Interaction of a PDZ protein domain with a synthetic library of all human protein C termini. Angew Chem Int Edit 38:2000–2004

102. Howell BW, Lanier LM, Frank R, Gertler FB, Cooper JA (1999) The disabled 1 phosphotyrosine-binding domain binds to the internalization signals of transmembrane glycoproteins and to phospholipids. Mol Cell Biol 19:5179–5188

103. Dostmann WRG, Nickl C, Thiel S, Tsigelny I, Frank R, Tegge WJ (1999) Delineation of selective cyclic GMP-dependent protein kinase I alpha substrate and inhibitor peptides based on combinatorial peptide libraries on paper. Pharmacol Ther 82:373–387

104. Brix J, Rüdiger S, Bukau B, Schneider-Mergener J, Pfanner N (1999) Distribution of binding sequences for the mitochondrial import receptors Tom20, Tom22, and Tom70 in a presequence-carrying preprotein and a non-cleavable preprotein. J Biol Chem 274:16522– 16530

105. Torrens I, Reyes O, Ojalvo AG, Seralena A, Chinea G, Cruz LJ, de la Fuente J (1999) Mapping of the antigenic regions of streptokinase in humans after streptokinase therapy. Biochem Biophys Res Commun 259:162–168

106. Welschof M, Reineke U, Kleist C, Kipriyanov S, Little M, Volkmer-Engert R, Schneider-Mergener J, Opelz G, Terness P (1999) The antigen binding domain of non-idiotypic human anti-F(ab')(2) autoantibodies: study of their interaction with IgG hinge region epitopes. Hum Immunol 60:282–290

107. Pullen SS, Dang TTA, Crute JJ, Kehry MR (1999) CD40 signaling through tumor necrosis factor receptor- associated factors (TRAFs) - binding site specificity and activation of downstream pathways by distinct TRAFs. J Biol Chem 274:14246–14254

108. Reinhard M, Zumbrunn J, Jaquemar D, Kuhn M, Walter U, Trueb B (1999) An alpha-actinin binding site of zyxin is essential for subcellular zyxin localization and alpha-actinin recruitment. J Biol Chem 274:13410–13418

109. Haan S, Hemmann U, Hassiepen U, Schaper F, Schneider-Mergener J, Wollmer A, Heinrich PC, Grotzinger J (1999) Characterization and binding specificity of the monomeric STAT3-SH2 domain. J Biol Chem 274:1342–1348

110. Liu ZH, Song DY, Kramer A, Martin ACR, Dandekar T, Schneider-Mergener J, Bautz EKF, Dubel S (1999) Fine mapping of the antigen-antibody interaction of scFv215, a recombinant antibody inhibiting RNA polymerase II from Drosophila melanogaster. J Mol Recognit 12:103–111

111. Hilpert K, Behlke J, Scholz C, Misselwitz R, Schneider-Mergener J, Höhne W (1999) Interaction of the capsid protein p24 (HIV-1) with sequence-derived peptides: influence on p24 dimerization. Virology 254:6–10

112. Reuter M, Schneider-Mergener D, Kupper D, Meisel A, Mackeldanz P, Kruger DH, Schroeder C (1999) Regions of endonuclease EcoRII involved in DNA target recognition identified by membrane-bound peptide repertoires. J Biol Chem 274:5213–5221

113. Valle M, Kremer L, Martínez C, Roncal F, Valpuesta JM, Albar JP, Carrascosa JL (1999) Domain architecture of the bacteriophage Φ29 connector protein. J Mol Biol 288:899–909

114. Retzer MD, Yu RH, Schryvers AB (1999) Identification of sequences in human transferrin that bind to the bacterial receptor protein, transferrin-binding protein B. Mol Microbiol 32:111–121

115. Valle M, Munoz M, Kremer L, Valpuesta JM, Martinez A-C, Carrascosa JL, Albar JP (1999) Selection of antibody probes to correlate protein sequence domains with their structural distribution. Protein Sci 8:883–889

116. Gonsior SM, Platz S, Buchmeier S, Scheer U, Jockusch BM, Hinssen H (1999) Conformational difference between nuclear and cytoplasmic actin as detected by a monoclonal antibody. J Cell Sci 112:797–809
117. Reineke U, Sabat R, Misselwitz R, Welfle H, Volk HD, Schneider-Mergener J (1999) A synthetic mimic of a discontinuous binding site on interleukin-10. Nat Biotechnol 17:271–275
118. Winkler K, Scharnagl H, Tisljar U, Hoschutzky H, Friedrich I, Hoffmann MM, Huttinger M, Wieland H, Marz W (1999) Competition of A beta amyloid peptide and apolipoprotein E for receptor-mediated endocytosis. J Lipid Res 40:447–455
119. Piossek C, Schneider-Mergener J, Schirner M, Vakalopoulou E, Germeroth L, Thierauch KH (1999) Vascular endothelial growth factor (VEGF) receptor II-derived peptides inhibit VEGF. J Biol Chem 274:5612–5619
120. Blüthner M, Schafer C, Schneider C, Bautz FA (1999) Identification of major linear epitopes on the sp100 nuclear PBC antoantigen by the gene-fragment phage-display technology. Autoimmunity 29:33–42
121. Monnet C, Laune D, Laroche-Traineau J, Biard-Piechaczyk M, Briant L, Bes C, Pugniere M, Mani JC, Pau B, Cerutti R, Devauchelle G, Devaux C, Granier C, Chardes T (1999) Synthetic peptides derived from the variable regions of an anti-CD4 monoclonal antibody bind to CD4 and inhibit HIV-1 promoter activation in virus-infected cells. J Biol Chem 274:3789–3796
122. Partidos CD, Salani FB, Ripley J, Steward MW (1999) Deconstructing the antigenic profile of a protective epitope from measles-virus fusion protein using overlapping peptides. Vaccine 18:321–324
123. Bittorf T, Sasse T, Wright M, Jaster R, Otte L, Schneider-Mergener J, Brock J (2000) cDNA cloning and functional analysis of a truncated STAT5a protein from autonomously growing FDCP-1 cells. Cell Signal 12:721–730
124. Hilpert K, Hansen G, Wessner H, Schneider-Mergener J, Höhne W (2000) Characterizing and optimizing protease/peptide inhibitor interactions, a new application for spot synthesis. J Biochem (Tokyo) 128:1051–1057
125. Winkler K, Kramer A, Kuttner G, Seifert M, Scholz C, Wessner H, Schneider-Mergener J, Hohne W (2000) Changing the antigen binding specificity by single point mutations of an anti-p24 (HIV-1) antibody. J Immunol 165:4505–4514
126. Siray H, Frommel C, Voronkova T, Hahn S, Arnold W, Schneider-Mergener J, Scherneck S, Ulrich R (2000) An immunodominant, cross-reactive B-cell epitope region is located at the C-terminal part of the hamster polyomavirus major capsid protein VP1. Viral Immunol 13:533–545
127. Dostmann WRG, Taylor MS, Nickl CK, Brayden JE, Frank R, Tegge WJ (2000) Highly specific, membrane-permeant peptide blockers of cGMP-dependent protein kinase I alpha inhibit NO-induced cerebral dilation. Proc Natl Acad Sci USA 97:14772–14777
128. Rüdiger S, Mayer MP, Schneider-Mergener J, Bukau B (2000) Modulation of substrate specificity of the DnaK chaperone by alteration of a hydrophobic arch. J Mol Biol 304:245–251
129. Mahler M, Mierau R, Blüthner M (2000) Fine-specificity of the anti-CENP-A B-cell autoimmune response. J Mol Med 78:460–467
130. Pistor S, Grobe L, Sechi AS, Domann E, Gerstel B, Machesky LM, Chakraborty T, Wehland J (2000) Mutations of arginine residues within

the 146-KKRRK-150 motif of the ActA protein of *Listeria monocytogenes* abolish intracellular motility by interfering with the recruitment of the Arp2/3 complex. J Cell Sci 113:3277–3287

131. Hoffmüller U, Knaute T, Hahn M, Höhne W, Schneider-Mergener J, Kramer A (2000) Evolutionary transition pathways for changing peptide ligand specificity and structure. EMBO J 19:4866–4874

132. Takahashi M, Ueno A, Mihara H (2000) Peptide design based on an antibody complementarity-determining region (CDR): construction of porphyrin-binding peptides and their affinity maturation by a combinatorial method. Chem Eur J 6:3196–3203

133. Ferrieres G, Pugniere M, Mani JC, Villard S, Laprade M, Doutre P, Pau B, Granier C (2000) Systematic mapping of regions of human cardiac troponin I involved in binding to cardiac troponin C: N- and C-terminal low affinity contributing regions. FEBS Lett 479:99–105

134. Torreilles F, Roquet F, Granier C, Pau B, Mourton-Gilles C (2000) Binding specificity of monoclonal antibody AD2: influence of the phosphorylation state of tau. Mol Brain Res 78:181–185

135. Drees B, Friederich E, Fradelizi J, Louvard D, Beckerle MC, Grolsteyn RM (2000) Characterization of the interaction between zyxin and members of the ena/vasodilator-stimulated phosphoprotein family of proteins. J Biol Chem 275:22503–22511

136. Jensen M, Hartmann T, Engvall B, Wang R, Uljon SN, Sennvik K, Naslund J, Muehlhauser F, Nordstedt C, Beyreuther K, Lannfelt L (2000) Quantification of Alzheimer amyloid beta peptides ending at residues 40 and 42 by novel ELISA systems. Mol Med 6:291–302

137. Osman AA, Günnel T, Dietl A, Uhlig HH, Amin M, Fleckenstein B, Richter T, Mothes T (2000) B cell epitopes of gliadin. Clin Exp Immunol 121:248–254

138. Joki-Korpela P, Roivainen M, Lankinen H, Poyry T, Hyypia T (2000) Antigenic properties of human parechovirus 1. J Gen Virol 81:1709–1718

139. Böldicke T, Struck F, Schaper F, Tegge W, Sobek H, Villbrandt B, Lankenau P, Bocher M (2000) A new peptide-affinity tag for the detection and affinity purification of recombinant proteins with a monoclonal antibody. J Immunol Methods 240:165–183

140. Hammerschmidt S, Tillig MP, Wolff S, Vaerman JP, Chhatwal GS (2000) Species-specific binding of human secretory component to SpsA protein of *Streptococcus pneumoniae* via a hexapeptide motif. Mol Microbiol 36:726–736

141. Laune D, Molina F, Mani JC, Del Rio M, Bouanani M, Pau B, Granier C (2000) Dissection of an antibody paratope into peptides discloses the idiotope recognized by the cognate anti-idiotypic antibody. J Immunol Methods 239:63–73

142. Carbonnet F, Blanchard D, Hattab C, Cochet S, Petit-Leroux Y, Loirat MJ, Cartron JP, Bertrand O (2000) A murine monoclonal antibody against Kx protein which reacts also with beta-spectrin. Transfus Med 10:145–154

143. Fu J, Hato M, Ohmae H, Matsuoka H, Kawabata M, Tanabe K, Miyamoto Y, Leafasia JL, Chinzei Y, Ohta N (2000) Epitope-specific impairment of production of antibody against merozoite surface glycoprotein 1 of *Plasmodium falciparum* in symptomatic patients with malaria. Parasitol Res 86:345–351

144. Basmaciogullari S, Autiero M, Culerrier R, Mani JC, Gaubin M, Mishal Z, Guardiola J, Granier C, Piatier-Tonneau D (2000) Mapping the CD4 binding domain of gp17, a glycoprotein secreted from seminal vesicles and breast carcinomas. Biochemistry 39:5332–5340

145. Ball LJ, Kühne R, Hoffmann B, Häfner A, Schmieder P, Volkmer-Engert R, Hof M, Wahl M, Schneider-Mergener J, Walter U, Oschkinat H, Jarchau T (2000) Dual epitope recognition by the VASP EVH1 domain modulatesa polyproline ligand specificity and binding affinity. EMBO J 19:4903–4914

146. Gouin-Charnet A, Laune D, Granier C, Mani JC, Pau B, Mourad G, Argiles A (2000) Alpha(2)-macroglobulin, the main serum antiprotease, binds beta(2)-microglobulin, the light chain of the class I major histocompatibility complex, which is involved in human disease. Clin Sci 98:427–433

147. Fu J, Hato M, Igarashi K, Suzuki T, Matsuoka H, Ishii A, Leafasia JL, Chinzei Y, Ohta N (2000) A simple screening method for detecting bindings between oligopeptides and HLA-DR molecules on filter papers: possible application for mapping of putative helper T-cell epitopes on MSP1 of *Plasmodium falciparum.* Microbiol Immunol 44:249–257

148. Tunback P, Liljeqvist JA, Lowhagen GB, Bergstrom T (2000) Glycoprotein G of herpes simplex virus type 1: identification of type-specific epitopes by human antibodies. J Gen Virol 81:1033–1040

149. Krause M, Sechi AS, Konradt M, Monner D, Gertler FB, Wehland J (2000) Fyn-binding protein (Fyb)/SLP-76-associated protein (SLAP) Ena/vasodilator-stimulated phosphoprotein (VASP) proteins and the Arp2/3 complex link T cell receptor (TCR) signaling to the actin cytoskeleton. J Cell Biol 149:181–194

150. Ferrieres G, Villard S, Pugniere M, Mani JC, Navarro-Teulon I, Rharbaoui F, Laune D, Loret E, Pau B, Granier C (2000) Affinity for the cognate monoclonal antibody of synthetic peptides derived from selection by phage display – role of sequences flanking the binding motif. Eur J Biochem 267:1819–1829

151. Blüthner M, Mahler M, Müller DB, Dünzl H, Bautz FA (2000) Identification of an alpha-helical epitope region on the PM/Scl-100 autoantigen with structural homology to a region on the heterochromatin p25 beta autoantigen using immobilized overlapping synthetic peptides. J Mol Med 78:47–54

152. Otvos L, Pease AM, Bokonyi K, Giles-Davis W, Rogers ME, Hintz PA, Hoffmann R, Ertl HCJ (2000) In situ stimulation of a T helper cell hybridoma with a cellulose-bound peptide antigen. J Immunol Methods 233:95–105

153. Amatschek K, Necina R, Hahn R, Schallaun E, Schwinn H, Josic D, Jungbauer A (2000) Affinity chromatography of human blood coagulation factor VIII on monoliths with peptides from a combinatorial library. HRC-J High Resolut Chromatogr 23:47–58

154. Loog M, Toomik R, Sak K, Muszynska G, Jarv J, Ek P (2000) Peptide phosphorylation by calcium-dependent protein kinase from maize seedlings. Eur J Biochem 267:337–343

155. Erck C, Frank R, Wehland J (2000) Tubulin-tyrosine ligase, a long-lasting enigma. Neurochem Res 25:5–10

156. Himpel S, Tegge W, Frank R, Leder S, Joost HG, Becker W (2000) Specificity determinants of substrate recognition by the protein kinase DYRK1A. J Biol Chem 275:2431–2438

157. Sachse K, Helbig JH, Lysnyansky I, Grajetzki C, Müller W, Jacobs E, Yogev D (2000) Epitope mapping of immunogenic and adhesive structures in repetitive domains of *Mycoplasma bovis* variable surface lipoproteins. Infect Immun 68:680–687
158. Rau HK, DeJonge N, Haehnel W (2000) Combinatorial synthesis of four-helix bundle hemoproteins for tuning of cofactor properties. Angew Chem Int Edit 39:250
159. Portefaix J-M, Thebault S, Bourgain-Guglielmetti F, Del Rio M, Granier C, Mani J-C, Navarro-Teulon I, Nicolas M, Soussi T, Pau B (2000) Critical residues of epitopes recognized by several anti-p53 monoclonal antibodies correspond to key residues of p53 involved in interaction with the mdm2 protein. J Immunol Methods 244:17–28
160. Maier B, Molinger M, Cope AP, Fugger L, Schneider-Mergener J, Sonderstrup G, Kamradt T, Kramer A (2000) Multiple cross-reactive self-ligands for *Borrelia burgdorferi*-specific HLA-DR4-restricted T cells. Eur J Immunol 30:448–457
161. Licha K, Bhargava S, Rheinländer C, Becker A, Schneider-Mergener J, Volkmer-Engert R (2000) Highly parallel nano-synthesis of cleavable peptide-dye conjugates on cellulose membranes. Tett Lett 41:1711- 1715
162. Dekker N, Cox RC, Kramer RA, Egmond MR (2001) Substrate specificity of the integral membrane protease OmpT determined by spatially addressed peptide libraries. Biochemistry 40:1694–1701
163. Geginat G, Schenk S, Skoberne M, Goebel W, Hof H (2001) A novel approach of direct ex vivo epitope mapping identifies dominant and subdominant CD4 and CD8 T cell epitopes from *Listeria monocytogenes*. J Immunol 166:1877–1884
164. Rüdiger S, Schneider-Mergener J, Bukau B (2001) Its substrate specificity characterizes the DnaJ co-chaperone as a scanning factor for the DnaK chaperone. EMBO J 20:1042–1050
165. Töpert F, Pires R, Landgraf C, Oschkinat H, Schneider-Mergener J (2001) Synthesis of an array comprising 837 variants of the hYAP WW protein domain. Angew Chem Int Edit Engl 40:897–900
166. Schnepf R, Hörth P, Bill E, Wieghardt K, Hildebrandt P, Haehnel W (2001) De novo design and characterization of copper centers in synthetic sour-helix-bundle proteins. J Am Chem Soc 123:2186–2195
167. Coulier L, Laune D, Orfanoudakis G, Wlad H, Janson JC, Granier C, Altschuh D (2001) Delineation of a linear epitope by multiple peptide synthesis and phage display. J Immunol Methods 249:253–264

References

Borman S (2000) Combinatorial synthesis hits the spot. Chem Eng News 78:25–27
Frank R (1994) SPOT-synthesis: an easy and flexible tool to study molecular recognition. In: Epton R (ed) Proc Int Symp on Innovation and Perspectives in Solid Phase Synthesis, Oxford, August 1993. Mayflower Worldwide Ltd, Birmingham, pp 509–512
Frank R, Güler S (1990) Verfahren zur schnellen Synthese von trägergebundenen oder freien Peptiden oder Oligonucleotiden, damit hergestelltes

Flachmaterial, dessen Verwendung sowie Vorrichtung zur Durchführung des Verfahrens. German Patent application P 40 27 657.9

Frank R, Güler S, Krause S, Lindenmaier W (1991) Facile and rapid spot-synthesis of large numbers of peptides on membrane sheets. In: Giralt E, Andreu D (eds) Peptides 1990. Proc. 21st Eur Peptide Symp. ESCOM Science Publishers BV, Leiden, pp 151–152

Gallop MA, Barrett RW, Dower WJ, Fodor SPA, Gordon EM (1994) Applications of combinatorial technologies to drug discovery. 1. Background and peptide combinatorial libraries. J Med Chem 37:1233–1251

Kramer A, Volkmer-Engert R, Malin R, Reineke U, Schneider-Mergener J (1993) Simultaneous synthesis of peptide libraries on single resin and continuous cellulose membrane supports: examples for the identification of protein, metal and DNA binding peptide mixtures. Peptides Res 6:314–319

Pirrung MC, Read JL, Fodor SPA, Stryer L (1990) Large scale photolithographic solid phase synthesis of polypeptides and receptor binding sceening thereof. United States Patent application 492,462

Southern EM (1988) Analyzing polynucleotide sequences. Int Patent application PCT GB 89/00460

Stockwell RB (2000) Chemical genetics: ligand-based discovery of gene function. Nat Rev 1:116–125

Chemistry of Fmoc Peptide Synthesis on Membranes

NORBERT ZANDER and HEINRICH GAUSEPOHL

Introduction

Peptide synthesis is a repetitive procedure schematically shown in Fig. 1. Each cycle of deprotection, wash, coupling and wash introduces one amino acid building block (residue) to the growing chains anchored covalently to an insoluble solid support via the carboxy terminus. Temporary protection for the α-amino group of the incoming residue, which is removed in each cycle (deprotection), ensures that the residue is coupled to the growing chain only once. Coupling requires prior activation of the carboxy group of the incoming residue, as explained in more detail below. To prevent undesired reactions, the side-chain functional groups are also protected during the synthesis and are deprotected together at the end of the synthesis. Between individual operations, the support has to be thoroughly washed in order to remove excess reagents from the growing chains. This is actually the most important feature of solid-phase peptide synthesis (SPPS) for which Bruce Merrifield was awarded the Nobel Prize in 1984. Typically, the peptides are cleaved from the support together with the side-chain protecting groups, but there are several experimental procedures, which require solid-phase-

N. Zander (✉) (e-mail: nza@aims-scientific-products.de,
Tel.: +49-531-2602865, Fax +49-531-2802866)
AIMS Scientific Products GmbH, Mascheroder Weg 1B, 38124 Braunschweig, Germany
H. Gausepohl
INTAVIS AG, Friedrich-Ebert-Strasse, 51429 Bergisch Gladbach, Germany

Springer Lab Manual
J. Koch, M. Mahler (Eds.) Peptide Arrays on Membrane Supports

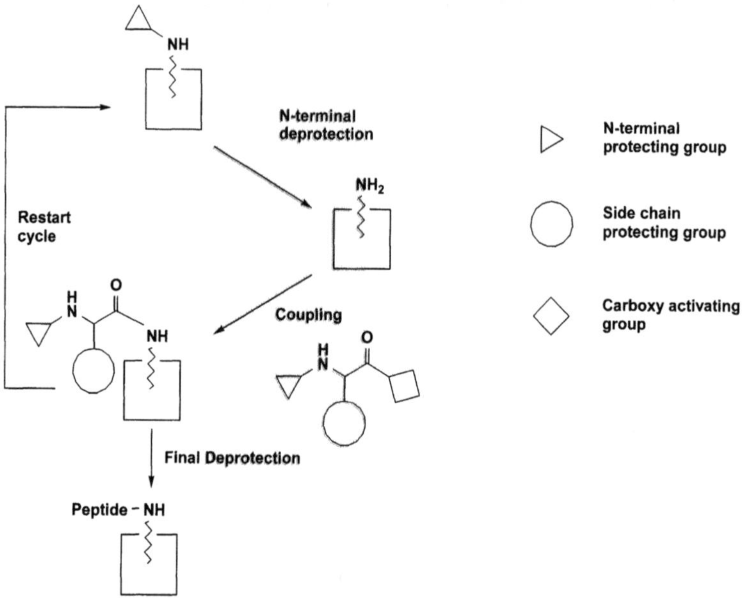

Fig. 1. Peptide synthesis on solid phase. Each cycle elongates the growing peptide chain by one amino acid residue and starts with removal of the N-terminal protecting group of the immobilized peptide chain followed by coupling of the activated amino acid derivative bearing an N-terminal and, if necessary, a side-chain protecting group. Finally, the peptide is generated by the removal of all side-chain protecting groups

bound peptides. The SPOT technique uses membranes as the solid support and enables the parallel synthesis and testing of hundreds to thousands of peptides at different locations on one membrane (Frank 1992).

The two major strategies in use today are named after the temporary protection used for the α-amino group: (1) Fmoc (=9-fluorenylmethyloxycarbonyl), and (2) BOC (=*tert*-butyloxycarbonyl). Their chemical structures and the cleavage conditions applied for their removal are shown in Fig. 2. The BOC-group is cleaved by acid in each cycle and a very strong acid-like hydrogen fluoride (HF) must be used for final cleavage of the side-chain protecting groups. The Fmoc group is cleaved by a mild base, usually piperidine, and a strong acid, trifluoroacetic acid (TFA), can be used for final cleavage. Fmoc chemistry is therefore a milder procedure and clearly preferable in manual protocols. Fmoc solid-phase peptide synthesis was reviewed recently (Chan and White 2000).

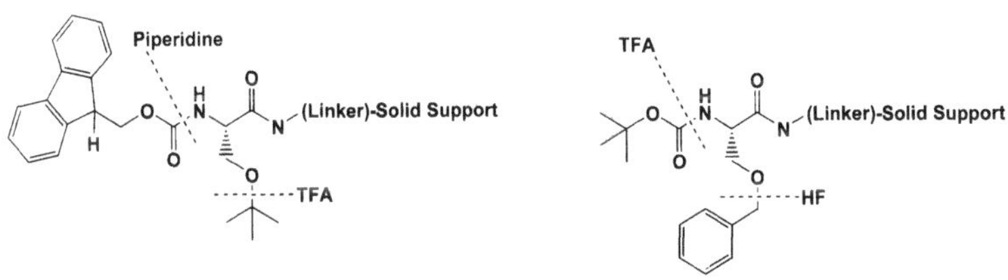

Fig. 2. Protecting group strategies using either Fmoc (*left*) or BOC (*right*) as the N-terminal protecting group. The indicated reagents are used to cleave the temporary N-terminal and the side-chain protecting groups

Deprotection of the N-terminal Fmoc protecting group

While the Fmoc group has been shown to be completely stable even to strong acids like TFA or HBr in acetic acid for 1–2 days, it is rapidly removed by base (Fields and Noble 1990). The cleavage is fast with primary and some secondary amines and slow with tertiary amines. The mechanism of this reaction is shown in Fig. 3. The lone hydrogen atom on the fluorene ring system in **1** is unusually acidic. The corresponding anion is aromatic and therefore stabilized. A deprotonation with a weak base, followed by a β-elimination, generates the dibenzofulvene **2**. The free N-terminal amino group **4** is liberated from the unstable N-substituted carbamic acid **3** by an irreversible loss of carbon dioxide. The final products of the deprotection with primary and secondary amines are the strongly UV absorbing dibenzofulvene adducts **5**. This allows the monitoring of the deprotection reaction by measuring the change in the optical density over time.

Typically, the Fmoc group is removed by treatment with 20–50 % (v/v) piperidine in dimethyl formamide (DMF) in a very fast reaction with a half life ($t_{1/2}$) of only 0.1 min. This reaction should be performed in a fume hood due to the unpleasant smell and the toxicity of piperidine.

The most commonly used solvents for SPPS on membranes are DMF and 1-methyl-2-pyrrolidone (NMP). Unfortunately, these solvents are known to contain dimethylamine or methylamine as basic impurities, which slowly cleave the Fmoc group. It is therefore advisable to monitor the amine content of these solvents. While high-quality DMF is available, the amine content of 1-methyl-2-pyrrolidone should be tested (see Procedure section, "Monitoring of the amine content").

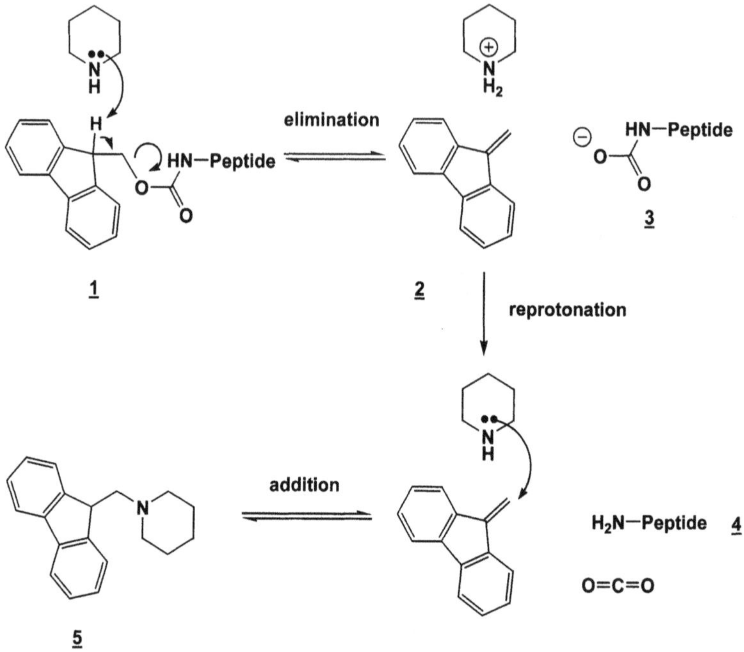

Fig. 3. Cleavage of the Fmoc group with piperidine. In the first step, piperidine eliminates the peptide as an N-substituted carbamic acid **3** from the dibenzofulvene **2** and adds to **2** after reprotonation to form the adduct **5**. The unstable carbamic acid **3** eliminates carbon dioxide and releases the free peptide **4**

Activation and coupling

After deprotection and washing, the support is ready for the next coupling step. To generate the amide bond with the free amino group on the growing peptide chain, the carboxy group of the incoming residue must be activated. There are several reagents to do this but the most suitable method for procedures within the scope of this book is more than 25 years old: Addition of diisopropylcarbodiimide (DIC) and 1-hydroxybenzotriazole (HOBt) to the Fmoc amino acid will generate the corresponding HOBt esters in situ, which react rapidly with the free amino groups of the growing peptide chains. The reactivity of amino and carboxy groups are the reason for a feature not yet mentioned: Amino acids with reactive groups on their side chains must be protected during the entire synthesis. A list of commonly used amino acid derivatives with their side-chain protecting groups is given in Table 1. The molecular weight and formula are identical for both L- and D-enantiomers of each amino acid. Some Fmoc amino acids are commercially available as hydrates only and can be used in SPPS without dehydration. The data given in Table 1 are based on the product information of Alexis Deutschland (Grünberg, Germany).

This set of derivatives usually works smoothly with the protocols listed in this book. It may be helpful, however, to know some of their characteristics for troubleshooting if impure products are obtained or if analysis by mass spectrometry indicates incomplete deprotection.

Side reactions

All undesired reactions are called side reactions. They usually occur at a low level but should be considered in order to be able to interpret unexpected results. Incomplete coupling as well as incomplete Fmoc deprotection lead to amino acid deletions, residues missing in the final peptide. Some sequences show such problems after about 8–10 residues. Changing the solvent, extending the Fmoc deprotection time, and double or triple coupling can help. Incomplete removal of piperidine or amines in the solvent can cause premature Fmoc cleavage during the coupling and, consequently, multiple insertion of a residue. Some side chains can also cause problems:

- Arg (Pbf) The 2,2,4,6,7-pentametyldihydrobenzofuran-5-sulfonyl group (Pbf) is cleaved slowly and will probably not be fully removed under the cleavage conditions compatible with, e.g. cellulose membranes. The anion generated upon cleavage of Pbf is quite stable and will attach to the indole ring of tryptophan. Fmoc-Trp(Boc)-OH should be used to prevent this.
- Asn (Trt) The trityl group (Trt) is easily cleaved except if Asn is the N-terminal residue. In this case full cleavage can take almost as long as for Arg (Pbf).
- Cys (Acm) This group is not cleaved by acid. Assays requiring a free sulfhydryl group of cysteine should not be

Table 1. Commonly used Fmoc amino acid derivatives: structure and properties

Fmoc amino acid	Symbol	Side-chain protecting group	Structure	Formula	Molecular weight
Alanine	Fmoc-Ala-OH x H_2O	-	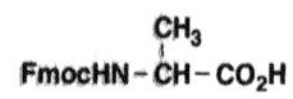	$C_{18}H_{17}NO_4$ x H_2O	329.3[a]
Arginine	Fmoc-Arg(Pbf)-OH	Pbf[b]	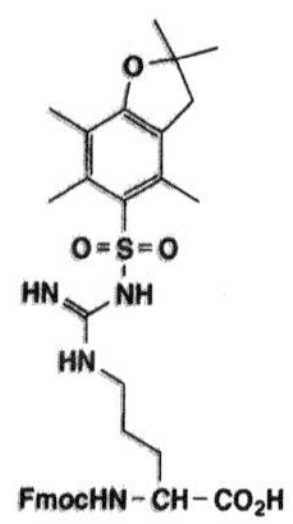	$C_{34}H_{40}N_4O_7S$	648.8
Asparagine	Fmoc-Asn(Trt)-OH x H_2O	Trt[c]	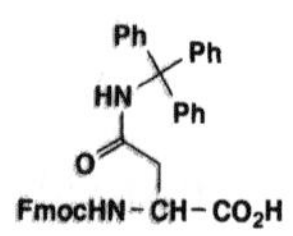	$C_{38}H_{32}N_2O_5$ x H_2O	614.7[a]
Aspartic acid	Fmoc-Asp(OtBu)-OH	OtBu[d]	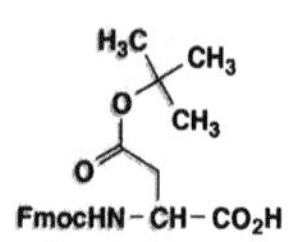	$C_{23}H_{25}NO_6$	411.5
Cysteine	Fmoc-Cys(Acm)-OH	Acm[e]	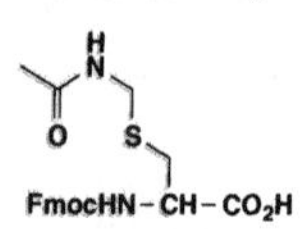	$C_{21}H_{22}N_2O_5S$	414.5
Glutamic acid	Fmoc-Glu(OtBu)-OH x H_2O	OtBu[d]	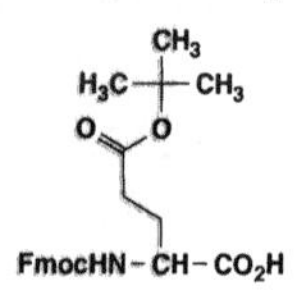	$C_{24}H_{27}NO_6$ x H_2O	443.5[a]
Glutamine	Fmoc-Gln(Trt)-OH	Trt[c]	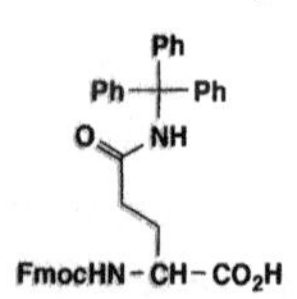	$C_{39}H_{34}N_2O_5$	610.7
Glycine	Fmoc-Gly-OH	-	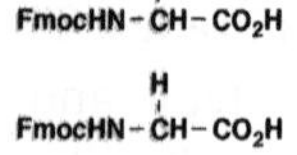	$C_{17}H_{15}NO_4$	297.3
Histidine	Fmoc-His(Boc)-OH	Boc	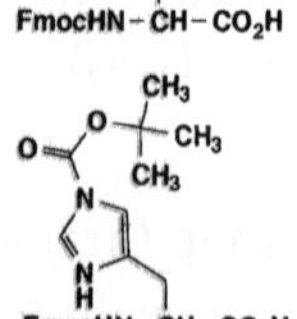	$C_{26}H_{27}N_3O_6$	477.5
Isoleucine	Fmoc-Ile-OH	-	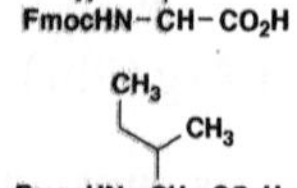	$C_{21}H_{23}NO_4$	353.4

Table 1. *(Continued)*

Fmoc amino acid	Symbol	Side-chain protecting group	Structure	Formula	Molecular weight
Leucine	Fmoc-Leu-OH	-	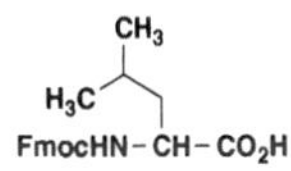	$C_{21}H_{23}NO_4$	353.4
Lysine	Fmoc-Lys(Boc)-OH	Boc	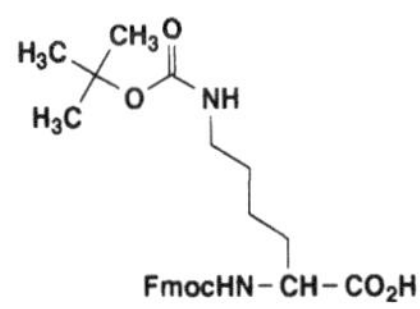	$C_{26}H_{32}N_2O_6$	468.5
Methionine	Fmoc-Met-OH	-	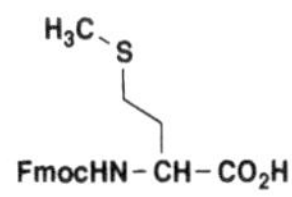	$C_{20}H_{21}NO_4S$	371.5
Phenyl-alanine	Fmoc-Phe-OH	-	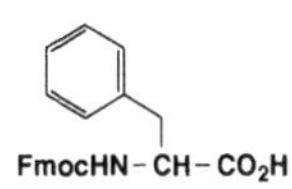	$C_{24}H_{21}NO_4$	387.4
Proline	Fmoc-Pro-OH x H_2O	-	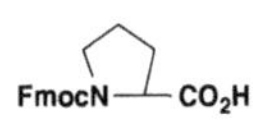	$C_{20}H_{19}NO_4$ x H_2O	355.4[a]
Serine	Fmoc-Ser(tBu)-OH	tBu[f]	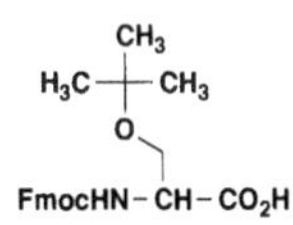	$C_{22}H_{25}NO_5$	383.4
Threonine	Fmoc-Thr(tBu)-OH	tBu[f]	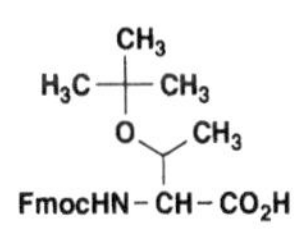	$C_{23}H_{27}NO_5$	397.5
Tryptophan	Fmoc-Trp(Boc)-OH	Boc	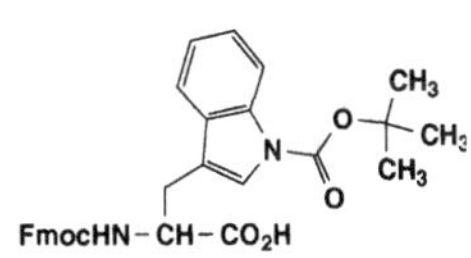	$C_{31}H_{30}N_2O_6$	526.6
Tyrosine	Fmoc-Tyr(tBu)-OH	tBu[f]	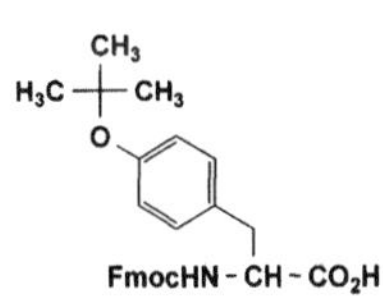	$C_{28}H_{29}NO_5$	459.6
Valine	Fmoc-Val-OH	-	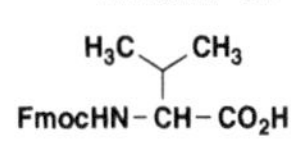	$C_{20}H_{21}NO_4$	339.4

[a] Including one molecule of water
[b] 2,2,4,6,7-Pentametyldihydrobenzofuran-5-sulfonyl (Pbf)
[c] trityl (Trt)
[d] *tert*-butyl-oxy (OtBu)
[e] Amidomethyl (Acm)
[f] *tert*-butyl (tBu)

attempted with membrane-based synthesis. Cysteine will oxidize immediately during handling and storage. If the undefined oxidized state is preferred over a small, remaining protection group, use the Fmoc-Cys(Trt)-OH derivative or the Fmoc-Cys(StBu)-OH and cleave the *tert*-butylthio group (StBu) with 1,4-dithio-DL-threitol (DTT) or mercapto-ethanol in water.

- Trp (Boc) Tryptophan is sensitive to protecting group fragments liberated by acid and should be protected by Boc. The Boc group is liberated upon aqueous work-up after cleavage.

Much more detailed information concerning the selection of certain side-chain protection groups and possible side reactions can be found in the protocol section of the Novabiochem-Calbiochem catalogue or in Fields and Noble (1990).

Membranes as solid support

There is a wide variety of solid supports and anchor groups used in peptide synthesis. As this chapter addresses only the synthesis of membrane-bound oligopeptides, the choice becomes much more limited. The membrane material plays a central role in the synthesis of immobilized peptides. The membrane must provide a functional group for the covalent attachment of the first amino acid, it has to be stable to the rather harsh synthesis conditions and, after synthesis, it must be compatible with the assay to be applied.

The most frequently used membrane material is cellulose (Frank and Overwin 1996). Esterification of the free hydroxy functions of the cellulose fibers with a Fmoc amino acid is a convenient method to introduce a spacer molecule and, after Fmoc deprotection, a free amino function for the SPPS of the peptide array. The high stability of the cellulose to organic solvents and bases allows the synthesis of peptides in excellent qualities by utilizing the standard Fmoc method. Furthermore, its hydrophilic structure is highly compatible with a wide variety of assay systems.

However, if the synthesized array is to be used repetitively, the limitations of the material must be considered. Difficulties could arise from the linkage of the peptides to the support via an ester bond. In aqueous basic solutions, not uncommon for assay and stripping conditions, the ester bond is cleaved and the peptides are removed from the membrane. Ninety percent of the peptides bound to the solid-phase via a glycine ester bond are hydrolyzed off overnight at room temperature in a phosphate buffer at pH 8.0. A new generation of commercially available membranes (see Materials) address this problem. The linkage of the spacer molecule, an aminated PEG derivative, to the membrane is stable to hydrolysis at pH 14 even overnight.

Standard cellulose and cellulose membranes show only limited acid stability. The commercially available aminated cellulose membranes are therefore made of especially acid-hardened cellulose paper in order to ensure mechanical stability after the deprotection of the acid-labile side-chain protecting groups with TFA. By following the stripping conditions (SDS, urea, acetic acid) described, these membranes can be reused 50 times and more (Frank and Owervin 1996). However, also with these membranes it is not advisable to use strongly acidic stripping conditions. The limited solvent compatibility of the cellulose membranes needs to be considered only if the peptides are to be modified during or after synthesis in solvents different from DMF, 1-methyl-2-pyrrolidone or water. Polar solvents are necessary to ensure a complete reaction while other common, less polar organic solvents like pyridine, THF or dichloromethane often lead to a low conversion. In order to overcome the chemical limitations of cellulose membranes, hydrophilic polyolefine membranes have been developed, e.g. by covalently grafting of a modified acrylamide layer onto a chemically inert polypropylene support. The commercially available membrane (see Materials) is stable to TFA, allows the use of a wide variety of solvents with different polarities, and is compatible with antibody-binding assays. The compatibility of this surface with other assays has yet to be established.

Derivatives of unusual amino acids

Besides the 20 natural amino acids, many organisms contain unnatural amino acids in some of their proteins, which were generated through post-translational modification. As the cells do this for a certain purpose, peptide segments containing these amino acids are often the most interesting ones in a protein, and there is a clear need to synthesize them. A wide variety of Fmoc-protected unnatural amino acid derivatives is available from various sources, of which some are listed at the end of this chapter. A more difficult situation arises if the modifications are phosphorylation, sulfation or glycosylation of certain amino acids (see Chap. 10). There are some commercially available derivatives but these should only be used with a sound knowledge of chemistry, as there are numerous side reactions. The protocol section of the Novabiochem-Calbiochem catalogue or Chan and White (2000) are good starting points for such experiments.

Synthesis of soluble peptides

The synthesis of soluble peptides enables testing in solution and quality control of the peptides synthesized. In the first step of the synthesis, the membrane has to be modified with a cleavable linker. After the attachment of the first amino acid to the linker, the synthesis of the peptide proceeds as described. The bond between linker and peptide must be stable during the synthesis and is cleaved after or during the side-chain deprotection step. An overview of useful commercially available linkers (see Materials) is given in Table 2.

The Fmoc Rink linker (Bernatowicz et al. 1989) and the Fmoc photo linker (Holmes and Jones 1995) can be used as the first residue in the synthesis just like any other Fmoc amino acid. The Rink linker is cleaved together with the side-chain protecting groups. The spots have to be separated before the TFA treatment. After evaporation of the deprotection mixture, the final products still contain the nonvolatile fragments of the Trt and Pbf protecting groups. The Fmoc photo linker is stable to TFA. The deprotection mixture and the side-chain group fragments can therefore be washed off prior to the linker cleavage. Potential problems could arise from partial degradation of tryptophan by the UV irradiation at 365 nm.

The synthesis of the diketopiperazine (DKP)-forming linker (Bray et al. 1990) starts with an Fmoc proline, esterified to the cellulose membrane. A membrane that is esterified uniformly

Table 2. Linker and cleavage conditions for the synthesis of soluble peptides

Linker	Structure	Cleavage conditions	C-terminal peptide modification
Rink linker	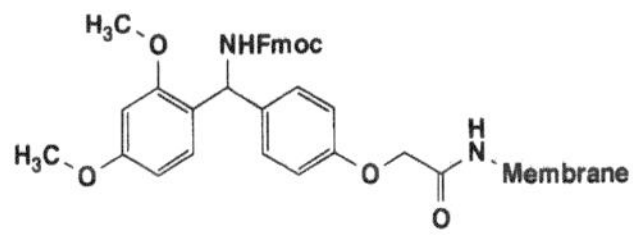	TFA	Carboxamide
Photo linker	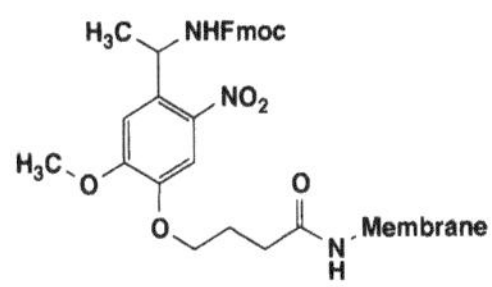	UV light at 365 nm	Carboxamide
Diketopiperzine - forming linker	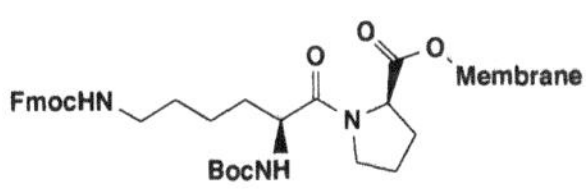	1. TFA 2. pH 7.5	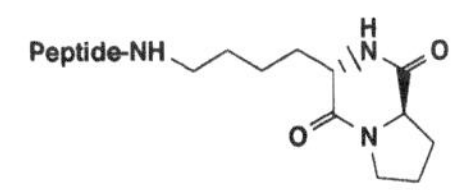
Estererification of the first amino acid	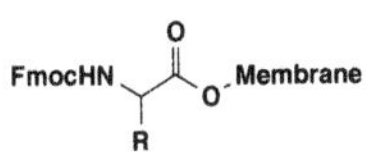	Aqueous NMe_3 Aqueous NH_3	Carboxylic Acid Carboxamide

with Fmoc proline is commercially available (see Materials). Peptide synthesis proceed as described after coupling of the unusual lysine derivative Boc-Lys(Fmoc)-OH. TFA deprotection of the side-chain protecting groups removes also the Boc group of the DKP-forming linker. TFA and the protecting group fragments can be washed off mildly acidic. Treatment with buffer at pH 7–7.5 cleaves the peptide from the membrane. All the peptides bear a C-terminal DKP group derived from lysine and proline.

Finally, direct esterification of the first amino acid to the cellulose membrane without any linker can be used for the synthesis of soluble peptides. The spot-wise esterification of Fmoc amino acids without a high degree of racemization is a demanding task and should be performed with a sound knowledge of chemistry only (Pilawa et al. 1999). Cellulose membranes that are esterified uniformly with, e.g. Fmoc glycine or Fmoc β-alanine are commercially available (see Materials) and can be used directly for the peptide synthesis. However, the additional C-terminal residue in all the peptides synthesized may influence functional tests. Following the side-chain deprotection step, the ester bond of all peptides on a membrane can be cleaved simultaneously without prior separation of the peptide spots with gaseous reagents. The peptides remain physically absorbed on the membrane and can be diluted directly into the assay buffer after separation of the spots. The product of the cleavage in an atmosphere over aqueous trimethylamine is the free carboxylic acid, while cleavage in an atmosphere over aqueous ammonia furnishes a mixture of the C-terminal carboxamide and the carboxylic acid in a ratio of roughly 4:1.

Materials

- DMF (peptide synthesis quality, Biosolve, 5554 HA Valkenswaard, NL)
- NMP (1-methyl-2-pyrrolidone absolute, Sigma-Aldrich Chemie, Deisenhofen, Germany)
- Acidic aluminum oxide (Aluminum oxide Fluka for chromatography Type 504C acidic, Sigma-Aldrich Chemie)
- Fmoc amino acids and linker (Alexis Deutschland, Grünberg, Germany; Bachem, Bubendorf, Switzerland; Calbiochem-Novabiochem, Bad Soden, Germany)
- Membranes (AIMS Scientific Products, Braunschweig, Germany)

Procedure

Note: Since DMF and 1-methyl-2-pyrrolidone are toxic, all steps should be carried out under a fume hood.

Monitoring of the amine content

Perform all tests in polyethylene or polypropylene tubes (no glass vials)!

1. Prepare a fresh stock solution of 1 % (w/v) bromophenol blue (BPB) dye in DMF.
2. Add 1 ml of the solvent to be tested into a 1.5-ml reaction tube and mix with 6 µl of the 1 % (w/v) BPB/DMF stock solution. The resulting color should be yellow for DMF or yellow to greenish yellow for 1-methyl-2-pyrrolidone. A blue color

indicates a high amine content. If the solvent is suspected to contain acids, which could result in a false negative test, add only 1 µl of the stock solution to 5 ml of the solvent. This time the mixture should turn light blue. A yellow color indicates the presence of acids and other impurities which might interfere with the synthesis and the BPB staining.

Purification of 1-methyl-2-pyrrolidone

1. Add 50 g of acidic aluminum oxide and 500 ml of absolute 1-methyl-2-pyrrolidone to a 1-l round-bottom flask, seal carefully, and shake vigorously on an orbital shaker overnight.
2. Let the aluminum oxide settle down, decant, and vacuum-filter through a 1 µm cellulose filter disc; this may take up to 1 h. Change the filter if necessary.

Synthesis of immobilized peptides on membranes

With the explanations outlined above, a general protocol can be derived:

1. Start with a membrane already derivatized with amino groups, a spacer molecule ending with an amino group or a Fmoc amino acid. A wide variety of membranes developed particularly for this purpose are commercially available from AIMS Scientific Products.
2. Cleave the Fmoc group off with a solution of 20% (v/v) piperidine in DMF.
3. Wash off excess piperidine with DMF.
4. Optional, but recommended: stain the free amino groups on the membrane with BPB, but only lightly. The staining allows monitoring of the piperidine removal. Traces of remaining piperidine cause a color change of the staining solution from yellow to blue. The amino groups on the membrane will stain only after complete removal of the piperidine.
5. Activate amino acid derivatives by addition of DIC and HOBt in 1-methyl-2-pyrrolidone.

6. Deliver aliquots of the solution to the membrane. This mode of activation does not require the addition of base. If the free amino groups on the membrane have been stained with BPB, the reaction can be followed visually. In the absence of base in the reaction solution, the membrane changes its color from blue, indicating free amino groups, to yellow when all amino functions have been acylated.

7. Repeat step 6 at least twice.

8. Wash thoroughly with a mixture of 2% (v/v) acetic anhydride in DMF. The addition of acetic anhydride to the DMF prevents premature cleavage of the Fmoc group by removing all traces of amine impurities from the DMF and acetylates unreacted amino functions on the membrane.

9. Wash off excess acetic anhydride with DMF

10. Start next cycle with step 2, or, after the last cycle, proceed with step 11.

End of the synthesis

11. Perform a final Fmoc deprotection (steps 2 and 3).

12. Wash off excess piperidine with DMF and stain with BPB (step 4).

13. Acetylate the terminal amino groups with a mixture of 2% (v/v) acetic anhydride in DMF (step 8).

14. Wash off excess acetic anhydride with DMF.

15. Wash membrane with ethanol and dry.

16. Cleave two times for 1 h with TFA/dichloromethane/water/diisobutylsilane 50:50:2:3 (v/v).

17. Wash with ethanol and dry

For detailed protocols, see Chapters 3 and 4.

Quality control

It is relatively easy to determine the quality of a synthetic peptide in solution, whereas it is quite difficult for peptides bound to a solid phase. A positive result in an assay can usually be trusted as there are numerous negative controls in the same experiment.

However, a negative result could be due to problems during the peptide synthesis. There are a few possibilities for quality control:

1. Monitor the presence and absence of free amino groups using the BPB test in each synthesis cycle. Store photocopies or electronic images for documentation.
2. Always include a few reference peptides for your assay on each membrane as positive controls.
3. If required, include a set of peptides starting with a cleavable linker in your synthesis. A suitable derivative is the Fmoc Rink linker (Novabiochem), which can be coupled just like an amino acid as the first residue on the support. Spots synthesized on this linker must be cut out of the membrane before the final side-chain deprotection step with TFA, as the linker is also cleaved under this condition. Cleave separately and analyze by high performance liquid chromatography (HPLC) and mass spectrometry (MS).

Synthesis of soluble peptides on membranes with the aid of the DKP-forming linker

According to the above, the general protocol for the synthesis of immobilized peptides can be modified as follows:

1. Start with a membrane already esterified with Fmoc proline (see Materials).
2. Start with Boc-Lys(Fmoc)-OH as the first residue and perform the peptide synthesis and side-chain deprotection as described above (steps 2–16).
3. Wash after TFA treatment with dichloromethane, 0.1 % HCl in 50 % aqueous methanol, and 1 M aqueous acetic acid
4. Dry in a desiccator over KOH.
5. Punch out the spots into 2-ml reaction tubes.
6. Eluate the peptides with 0.5 ml of a mixture of aqueous 0.1 M triethylamine and 20 % (v/v) ethanol (pH 7.0–7.5) at 30 °C overnight.

7. Concentrate in a vacuum centrifuge at room temperature and store at –20 °C.

References

Bernatowicz MS, Daniels, SB, Köster H (1989) A comparison of acid labile linkage agents for the synthesis of peptide C-terminal amides. Tetrahedron Lett 30:4645–4648

Bray AM, Maeji NJ, Geysen HM (1990) The simultaneous multiple production of solution phase peptides; assessment of the Geysen method of simultaneous peptide synthesis. Tetrahedron Lett 31:5811–5814

Chan WC, White PD (2000) Fmoc solid phase peptide synthesis: a practical approach. Oxford University Press, Oxford

Fields GB, Noble RL (1990) Solid phase peptide synthesis utilising 9-fluorenylmethoxycarbonyl amino acids. Int J Peptide Protein Res 35:161–214

Frank R (1992) Spot synthesis: an easy technique for the positionally addressable, parallel chemical synthesis on a membrane support. Tetrahedron 48:9217–9232

Frank R, Overwin H (1996) Spot synthesis. In: Morris GE (ed) Epitope mapping protocols. Methods in molecular biology. Humana, Totowa, New Jersey, pp 149–169

Holmes CP, Jones DG (1995) Reagents for combinatorial organic synthesis: development of a new o-nitrobenzyl photolabile linker for solid-phase synthesis. J Org Chem 60:2318–2319

Pilawa S, Zander N, Frank R (2001) Optimized reaction conditions for the direct esterification of protected amino acids to cellulose membrane supports by the SPOT technique. In: Epton R (ed) Innovation and perspectives in solid phase synthesis and combinatorial libraries. Proc 6th Int Symp, York, 1999. Mayflower Worldwide, Birmingham, UK, pp 337–338

Suppliers

Alexis Deutschland GmbH
(Tel.: +49-6401-900770, Fax: +49-6401-3823)
Giessener Strasse 12, 35305 Grünberg, Germany

Bachem AG
(e-mail: product information: documentation@bachem.com,
general information: marketing@bachem.ch,
complaints: sales.ch@bachem.com,
Tel.: +41-61-9312333, Fax: +41-61-9312549)
Hauptstrasse 144, 4416 Bubendorf, Switzerland

Calbiochem-Novabiochem GmbH
(e-mail: customer.service@calbiochem-novabiochem.de,
Internet: http://www.calbiochem-novabiochem.de/,
Tel.: +49-61-9663955, Fax: +49-61-9662361)
Postfach 1167, Ober der Roeth 4, 65812 Bad Soden,
65824 Schwalbach/Ts., Germany

AIMS Scientific Products GmbH
(e-mail: contact@aims-scientific-products.de,
Internet: http://www.aims-scientific-products.de,
Tel: +49-531-260-2865 or +49-177-7637299,
Fax: +49-531-260-2866)
Mascheroder Weg 1B, 38124 Braunschweig, Germany

SIGMA-ALDRICH Chemie GmbH
(Internet: http://www.sigma-aldrich.com)
Tel: +49-800-5155000
Postfach 1161, 82018 Taufkirchen

Biosolve BV (e-mail: biosolve@tip.nl,
Internet: http://www.biosolve.nl/NAW.html,
Tel: +31-40-2089322, Fax: +31-40-2048537)
Waalreseweg 17, 5554 HA Valkenswaard, The Netherlands

Manual Peptide Synthesis

GABRIELE PETERSEN

Introduction

Bruce Merrifield was awarded the Nobel Price in 1985 for the development of chemical peptide synthesis on solid supports (SPPS, solid-phase peptide synthesis), a technique, which, ever since, has tremendously advanced research in the fields of chemistry, biochemistry, molecular biology and medicine.

Ronald Frank's modification of peptide synthesis, the SPOT method (Frank 1992), allows for the synthesis of oligopeptide arrays on activated cellulose membranes, thus, using standard laboratory equipment, it is possible to synthesize a large amount of different membrane-bound peptides at relatively low cost. Although the method has been automated to synthesize several thousand peptides in a single run for high-throughput screening, various applications (discussed in other chapters of this manual) do not depend on a pipetting robot and can be carried out manually.

Outline

The basic procedure is outlined below (see also Fig. 1):

1. Design the peptides to be synthesized.
2. Dissolve Fmoc-protected activated amino acid esters in 1-methyl-2-pyrrolidone (NMP).

G. Petersen (✉) (e-mail: gabriele.petersen@urz.uni-heidelberg.de,
Tel: +49-6221-545256, Fax: +49-6221-545678)
Institute of Molecular Genetics, University of Heidelberg,
Im Neuenheimer Feld 230, 69120 Heidelberg, Germany

Springer Lab Manual
J. Koch, M. Mahler (Eds.) Peptide Arrays on Membrane Supports

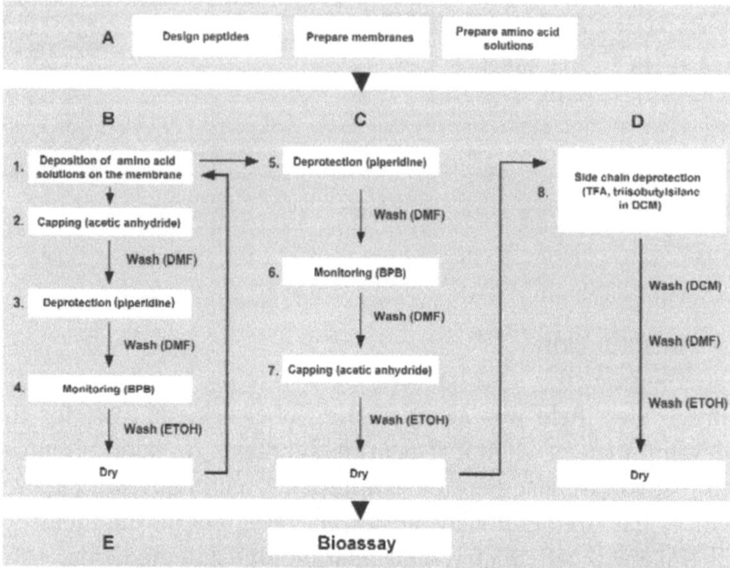

Fig. 1A–E. Outline of the manual peptide synthesis on activated membranes. **A** Preparatory steps, **B** repetitive synthesis cycles, **C** last synthesis cycle, **D** side-chain deprotection, **E** bioassay (see Chaps. 5–10)

3. Start synthesis cycles by depositing the first amino acid (note that chemical synthesis begins with the C-terminally located amino acid) on an activated membrane.
4. Cap unreacted amino groups with acetic anhydride.
5. Wash with dimethylformamide (DMF).
6. Remove the protective Fmoc group with piperidine.
7. Wash with DMF.
8. Stain for free amino groups with bromophenol blue (BPB).
9. Wash with ethanol.
10. Dry the membrane.
11. Repeat steps 3–10 according to the desired length of the oligopeptide.
12. Following the last cycle, remove protective groups from the amino acid side chains with triisobutylsilane in dichloromethane (DCM)/trifluoroacetic acid (TFA).

13. Wash in DCM.

14. Wash in DMF.

15. Wash in ethanol.

Materials

Equipment

General laboratory equipment such as fume hood, shaking platform, set of pipetmen, glass pipettes, DMF-resistant dispenser, etc.

You will also need a blow dryer which has a "cold" setting.

Follow the safety regulations and remember that you are working with hazardous chemicals, which cannot only harm your health: droplets of DMF, for example, may easily dissolve your pipetman.

Synthesis membranes

Ready-to-use membranes for the manual peptide synthesis are commercially available (Genosys Biotechnologies, London, UK) containing a grid of 8 × 12 distinct blue spots that represent free amino functions as visualized by the indicator bromophenol blue. These amino functions are usually provided by β-alanines as dipeptides or coupled to a polyethylene glycol spacer for better accessibility of the immobilized peptides (Frank and Overwin 1996).

Alternatively, you can buy any membrane containing a homogeneous surface of free amino functions (AIMS, ABIMED Analysen-Technik, Gemany) if you want to design your own grid. The grid is generated by depositing β-alanines or the first C-terminal amino acid of each peptide at the desired position. The volume applied in this first step should be about 30 % below the volume spotted in the subsequent synthesis cycles to ensure full-length peptides also at the borders of the spots.

Genosys also offers a SPOT Kit, including a membrane, 20 × 500 mg of Fmoc-L-amino acid active esters (Fmoc-Opfp-esters), 1 synthesis trough, 1 pair of forceps, 1 polypropylene box for final Fmoc deprotection, a manual and software. The kit greatly facilitates the synthesis setup.

Fmoc-L-amino acid active esters make it possible to simply dissolve the amino acids in 1-methyl-2-pyrrolidone without performing further activation steps (see Chap. 2). The synthesis trough, forceps, and polypropylene box are resistant to all chemicals used. Following installation of the software, you enter the sequence of the peptides to be synthesized; the software separates the various cycles and indicates the amino acids to be deposited at every spot. You only have to calculate the number of spots to be charged with the same amino acid and dissolve the according amount (with a 10 % extra for pipetting errors).

Chemicals

(For 1 membrane of 96 10-mer oligopeptides)

In some countries, purchase of the various chemicals requires a special permit, which is easily obtained by filling out a form usually provided by the supplier. You will have to specify what the chemicals will be used for, i. e. membrane-bound peptide synthesis, and that they will not be used for military and illegal purposes or for the treatment of humans and animals, etc.

Suppliers, e.g. Sigma-Aldrich, Fluka, Merck

- 1-Methyl-2 pyrrolidone (purest quality <100 ml)
- Molecular sieve 4 (<100 g)
- Dimethylformamide (2.5 l)
  For large amounts we recommend to buy DMF from SDS (Solvents, Documentation, Synthesis, France), which provides excellent quality at a very low price.
- Acetic anhydride (<100 ml)
- Piperidine (<100 ml)
- Bromophenol blue (<1 g)
- Ethanol (p.a. 250 ml)
- Dichloromethane (250 ml)
- Trifluoroacetic acid (<100 ml)
- Triisobutylsilane (<10 ml)

Amino Acids

Fmoc-protected amino acid active esters can be purchased from, e. g. Genosys or ABIMED.

Solutions

Quality control for DMF and 1-methyl-2-pyrrolidone

To ensure proper peptide synthesis, DMF and 1-methyl-2-pyrrolidone have to be tested for purity, i. e. the absence of free amino groups.

One mg of bromophenol blue dissolved in 10 ml DMF (0.01 % BPB in DMF) should yield an intense yellow to orange color. Dependent on the amount of free amino groups, the solution turns green or blue and another quality of DMF (synthesis grade for peptides) needs to be used.

1-methyl-2-pyrrolidone, however, can be cleaned by the addition of molecular sieve beads. Add 10 µl of a 1 % solution of BPB to 1 ml of 1-methyl-2-pyrrolidone. As for DMF, the solution should remain yellow and the presence of free amino groups is indicated by a color shift toward green or blue. To remove free amines, add molecular sieve beads (5 % of the liquid volume), mix, and let sit for a few days.

Amino acid solutions

Fmoc amino acid active esters should be dissolved in 1-methyl-2-pyrrolidone to yield 0.4 mM solutions. Per cycle you will need 2×0.9 µl of the respective amino acid solution per spot. Due to the multiple pipetting steps with small volumes and the viscous character of some of the amino acids, it is advisable to weigh in an extra 10–20 %. Since these derivatives are less reactive and less stable than freshly activated Fmoc amino acids (see Chaps. 2 and 4), solutions should be prepared shortly before use and stored in aliquots at –20 °C. Except for arginine, which is highly unstable, all other amino acids can be dissolved in one batch and aliquots should be prepared according to the amount needed per day. Aliquots of arginine powder should also be stored at –20 °C and dissolved in the required amount of 1-methyl-2-pyrrolidone before each synthesis cycle.

Capping

Four percent acetic anhydride (0.8 ml) in DMF (20 ml), freshly prepared before use (3× per cycle) and 2 % acetic anhydride (0.8 ml) in DMF (20 ml), freshly prepared before use (1×) for the last cycle

Staining

- 1 % (w/v) BPB in DMF (200 µl per cycle)

Removal of Fmoc groups

- 20 % (w/v) piperidine in DMF (20 ml per cycle)

Deprotection of side chains as final step

Prepare a fresh solution of dichloromethane (5 ml), trifluoro-acetic acid (250 µl) and triisobutylsilane (5 ml).

TFA is a very strong corrosive acid exuding harmful vapors and you must be extremely cautious. Always work in a fume hood and wear protective clothing including gloves and safety glasses. Never mix DMF and TFA (strong exothermic reaction). The reaction should be carried out in a polypropylene box with a lid (in the fume hood) and membranes have to be completely dry.

Procedure

Preparation of the membrane

A ready-to-use membrane has the size of a 96-well ELISA plate and contains a grid of 8×12 easily visible, blue spots with already deprotected free amino groups ready to receive the first amino acid of your peptide (remember that synthesis starts with the C-terminus). If you do not wish to use the complete membrane, simply cut the portion you need and freeze the remainder. Peptide synthesis is carried out in the absence of water, thus wetting of the membrane should be avoided and membranes as well as amino acid aliquots should be thawed in a desiccator or vacuum chamber.

The next step is to assign numbers to your blue spots. Although each cycle involves a staining step, which is supposed to stain free amino groups, this is not always the case and some spots simply cannot be stained. Depending on the amino acid sequence of your oligopeptide, some spots may vanish already after one or two cycles and never appear again. There is no reason to worry, even if you do not see anything, free amino groups will be present and you can proceed with the subsequent cycles. Some spots may even become stainable again upon deposition of the next amino acid.

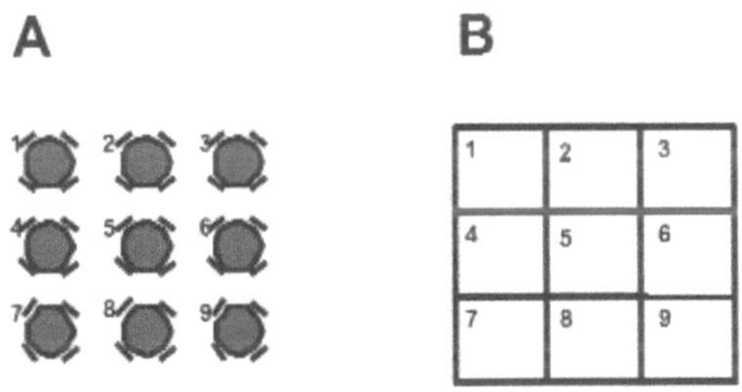

Fig. 2. A Spots numbered and marked on membranes with preformed grid; **B** screen for uniformly activated membranes

To determine the position of the solution to be deposited even if a spot becomes invisible upon synthesis, number the spots on an as yet unused membrane with a light graphite pencil and mark their outline as shown in Fig. 2A.

If you use a membrane without a preformed grid, prepare your own numbered screen with a light pencil according to Fig. 2B.

The active surface is then generated by spotting β-alanines or the first C-terminal amino acid of each peptide at the desired position within the frame. As mentioned above, the volume applied in this step should be about 30 % less than the volume spotted in the subsequent synthesis cycles to ensure full-length peptides also at the borders. Make sure that the squares are big enough to accept the applied volume.

Cut Whatman filter papers (according to the number of cycles to be performed) of approximately the same size as the membrane. Place the prepared membrane on one of the filter papers in a polypropylene synthesis tray in the fume hood. The paper will absorb excess liquid and should be replaced after each cycle to avoid cross-contamination.

Synthesis cycles

Spotting the amino acids

1. Thaw the amino acid solutions for the day in a desiccator and dissolve arginine (if needed in the cycle) in the required amount of 1-methyl-2-pyrrolidone. Start the synthesis by depositing 0.9 μl of the desired amino acid solution on the designated spot. Sort the amino acids alphabetically and

start with alanine if present in your sequence. Since you will probably synthesize different peptides, you should have a list (handmade or provided by the software which comes with the SPOT synthesis kit from Genosys) indicating which spot will need which amino acid. It is advisable to touch the membrane with the pipette tip since some of the amino acid solutions are rather viscous and you need the capillary force of the membrane to completely empty the tip. Once a spot has received its amino acid, mark it off on your pipetting list. Continue until all spots have received their first amino acid. The coupling reaction starts upon addition of the amino acid solution, and within 15 min of incubation time an efficiency of about 95% can be reached (Fields and Noble 1990). To increase the coupling efficiency, a second deposition of another 0.9 μl of the same amino acid is recommended. Since this step is rather tedious and time-consuming, it is almost certain that 15 min have passed when the last amino acid droplet of the first round has been deposited on the last spot and you can start with the second round of the first cycle. Incubate the membrane for another 15 min after the last drop of the second round has been placed before you start capping. Upon deposition of serine or threonine, blue spots may turn light green or yellow whereas lysine triggers a color change towards turquoise.

Capping

2. Although the coupling efficiency after two rounds of amino acid deposition is relatively high, the remaining free amino groups should be acetylated (capped) with acetic anhydride to avoid the synthesis of undesired peptides on a spot even if the percentage is low.
Remove the membrane from the synthesis tray and place it into a glass tray positioned on a slow-moving rocking platform ("seesaw"). Incubate the membrane three times with a mix of 20 ml of DMF+0.8 ml acetic anhydride for 30 s, 2 min and 5 min. Excess acetic anhydride is removed by 3 washes in 20 ml DMF each for 2 min.

Removal of protective Fmoc groups for the next synthesis cycle

3. Before starting the next synthesis cycle, Fmoc groups are removed by incubation with 20% piperidine in DMF. Following this 5-min incubation step, extensive washing with DMF is necessary to remove all traces of piperidine. Wash at least 10 times for 2 min each with 20 ml of DMF.

Monitoring free amino groups

4. Although staining of free amino groups with BPB is not guaranteed to be successful for every spot, it is still advisable to perform this step. Since the peptides remain membrane-bound for further use, no quality control can be performed, and even if some spots cannot be detected, BPB staining of the others assures that coupling and synthesis are performed properly. Incubate the membrane for 5 min in 0.01 % BPB in DMF (20 ml DMF with 0.2 ml 1 % BPB in DMF). This will not only stain the peptide spots, the entire membrane will turn light blue. Upon removal of excess BPB in ethanol (3×2 min) the membrane will revert to its white color and only the (stainable) spots will be visible. Dry the membrane wrapped in Whatman paper with a blow dryer (cold air setting). You can either proceed with the next synthesis cycle or freeze the membrane in a Ziploc plastic bag at –20 °C.

   Repeat steps 1–4 until the desired peptide is synthesized. You can easily synthesize peptides of differing lengths on the same membrane. Note that "complete" spots on a membrane which do not receive amino acids in the next cycle will be acetylated at the N-terminus unless you omit the capping step in the remaining cycles.

Last synthesis cycle

When you have deposited the amino acids of the second round of the last synthesis cycle (step 1), your assembled polypeptides will contain:

- Fmoc groups at the N-terminus
- Acetyl groups at the N-terminus of incomplete peptides if the coupling efficiency was low, and
- Various protective groups on the different side chains

5. After the required incubation time you should place the membrane for 5 min in 20 ml DMF and proceed with step 3 (removal of protective Fmoc groups), resulting in free amino groups at the N-termini of your peptides.

6. Continue with step 4 (Monitoring).

7. If you want to have acetyl groups present at the N-termini of the final peptides go to step 2 (Capping, spots will destain). Complete the last cycle by washing 3×2 min in 20 ml ethanol.

8. Following the last synthesis cycle, protective groups from the side chains have to be removed before the peptides can be used for their respective bioassays. If this step is performed immediately following the last synthesis cycle, place the membrane into the lidded polypropylene box and incubate for 1 h with a freshly prepared solution of dichloromethane (5 ml), TFA (250 µl), and triisobutylsilane (5 ml).

Note: If you have stored the membrane overnight in the freezer, it is advisable to thaw it in a desiccator, followed by a wetting step in ethanol with subsequent blow drying before you perform the deprotection step. Remember to obey the safety instructions regarding the handling of TFA! Wash the membrane in 20 ml each of dichloromethane (4×2 min), DMF (3×2 min), and ethanol (3×2 min). Dry the membrane with cold air.

Synthesis is now complete (depending on the amino acid sequence, some of the peptides will appear slightly yellow) and you can perform your desired assay (remember that the synthesis is performed in organic solvents, so you will have to hydrate the membrane in appropriate solutions (see Chaps. 5–10).

Store the membranes in a cold and dry place.

Short protocol

- Thaw membranes in vacuum desiccator
- Mark and number spots
- Thaw amino acids in desiccator
- Dissolve arginine in 1-methyl-2-pyrrolidone

1. Spot 0.9 µl amino acids
   Incubate 15 min

2. Spot 0.9 µl amino acids
   Incubate 15 min

3. 30 s 20 ml DMF+0.8 ml acetic anhydride
   2 min 20 ml DMF+0.8 ml acetic anhydride
   5 min 20 ml DMF+0.8 ml acetic anhydride
   2 min 20 ml DMF
   2 min 20 ml DMF
   2 min 20 ml DMF

4. 5 min 20 ml 20 % piperidine in DMF
5. Washes with 20 ml DMF
   10×2 min
6. 5 min DMF+0.2 ml 1 % BPB in DMF
7. Washes with 20 ml ethanol
   3×2 min
8. Dry the membrane

Continue with step 1 or freeze at –20 °C

Last cycle

1. Spot 0.9 μl amino acids
   Incubate 15 min
2. Spot 0.9 μl amino acids
   Incubate 15 min
3. 2 min 20 ml DMF
4. 5 min 20 ml 20 % piperidine in DMF
5. Washes with 20 ml DMF
   10×2 min
6. 5 min DMF+0.2 ml 1 %BPB in DMF
7. 20 ml DMF+0.4 ml acetic anhydride
   Incubate until all spots are destained
8. Washes with 20 ml DMF
   3×2 min
9. Washes with 20 ml ethanol
   3×2 min
   Deprotect side chains or freeze at –20 °C
   - Thaw frozen membrane in vacuum desiccator
   - Wash in ethanol
   - Dry

Side-chain deprotection

1. Incubate 1 h in 5 ml dichlormethane, 0.25 ml trifluoroacetic acid, 5 ml triisobutylsilane
2. Washes with 20 ml dichloromethane
   4×2 min
3. Washes with 20 ml DMF
   3×2 min

4. Washes with 20 ml ethanol
   3×2 min

5. Dry

Troubleshooting

- No staining with BPB

→ Deprotection or removal of piperidine not efficient

Comments

- Time considerations
  The synthesis of 8×12 10- to 15-mer peptides (=one membrane) requires about 1 week. You will need approximately 1 day to design the setup, calculate the amount of amino acids, prepare solutions (weighing of small amounts of amino acids especially arginine is relatively time-consuming). Depending on the number of peptides to be synthesized, deposition of 2×0.9 µl takes about 45 min and a complete cycle including all washing steps will require approximately 2 h. Three to four cycles a day are easily done with most of the time spent on extremely boring washing steps; however, deposition of the amino acid solutions requires your full attention, once the reaction has started there is no way back.

- Organic waste
  Peptide synthesis is performed in organic solvents, some of which are highly toxic. Due to the extensive washing steps, you will accumulate large volumes of organic waste. Check with your safety officer for appropriate waste containers.

- Manual synthesis of soluble peptides
  Soluble peptides can also be synthesized manually.

References

Fields GB, Noble RL (1990) Solid phase peptide synthesis utilising 9-fluorenylmethoxycarbonyl amino acids. Int J Peptide Protein Res 35:161–214

Frank R (1992) SPOT synthesis: an easy technique for the positionally addressable, parallel chemical synthesis on a membrane support. Tetrahedron 48:9217–9232

Frank R, Overwin H (1996) SPOT synthesis. In: Morris GE (ed) Epitope mapping protocols. Methods in molecular biology. Humana, Totowa, New Jersey, pp 149–169

Suppliers

ABIMED Analysen-Technik GmbH
(Tel.: +49-21-7389050, Fax: +49-21-73890577)
Raiffeisenstrasse 3, 40764 Langenfeld, Germany

Sigma-Genosys Ltd. (e-mail: info@sigma-genosys.co.uk
Internet: http://www.Genosys.co.uk/,
Tel.: +44-1223-839000, Fax: +44-1223-839300)
UK/Europe, London Road, Pampisford, Cambridge CB2 4EF, UK

Automated Synthesis of Solid-Phase Bound Peptides

HEINRICH GAUSEPOHL and CHRISTIAN BEHN

Introduction

The SPOT method was developed by Ronald Frank for simultaneous multiple peptide synthesis on separate sites on a homogeneous membrane carrier (Frank 1992). The principle of the technique is to dispense small droplets of pre-activated amino acid derivatives onto a predefined array of positions on a porous membrane. The droplets are absorbed and form individual reaction compartments for chemical reaction in solid-phase synthesis. A great number of distinct spots can be arranged on a large membrane sheet and each of these is individually addressable by manual or automated delivery of the respective reagent solution. The absorptive capacity of the membrane and the volume of the drops determine the number of spots per area and the spot size. According to the specific functionality of the matrix, the spot size correlates with the particular scale of the synthesis.

The synthesis is carried out using 9-fluorenyl-methoxycarbonyl (Fmoc)-protection chemistry on membranes made of pure cellulose. The amino acid building blocks are derivatives that are protected at their amino terminus by Fmoc. The protection group ensures that in each step only a single building block is coupled to each growing peptide chain. Trifunctional amino acids also carry a side-chain protection group. The Fmoc group must be removed in each synthesis cycle whereas the side-chain protection groups

H. Gausepohl
INTAVIS AG, Friedrich-Ebert-Strasse, 51429 Bergisch Gladbach, Germany
C. Behn (✉) (e-mail: behn@abimed.de,
Tel.: +49-2173-89050, Fax: +49-2173-890577)
ABIMED Analysen-Technik GmbH, Langenfeld, Germany

Springer Lab Manual
J. Koch, M. Mahler (Eds.) Peptide Arrays on Membrane Supports

are taken off at the end of the synthesis. In situ activation of the amino acid derivatives is performed by DIC/ HOBt, which leads to rapid activation and coupling. A membrane can be used to synthesize a large number of individual peptides, and these can be screened for biological activity by a Western-blot-like assay. Alternatively, the spots are cut out and cleaved separately from the support if a suitable linker was introduced prior to the synthesis.

The SPOT method was originally developed as a manual method and steps common to all spot reactors are carried out manually by washing the whole membrane with respective reagents and solvents (see Chap. 3). The most tedious step is the spotting of individual activated amino acids onto the reactive areas in patterns, which change in each cycle. Automation of this step has made the method applicable to a much wider range of experiments, as it cuts down the amount of time involved and allows the creation of much larger arrays with smaller volumes than can be handled manually. It is now possible to synthesize several thousand peptides in a single run. Automated deposition of amino acids also allows for easy double or triple couplings, which improves the synthesis quality.

Scope of automated SPOT synthesis

The ASP 222 automated spot robot was developed to reliably deliver amino acid derivatives to synthesize arrays on membranes. A pipetting robot under the control of PC-based software delivers volumes down to 100 nl to several thousand individual spots (Fig. 1). Four membranes of microtiter-plate format can be mounted on the work area. Standard grids have 96 or 384 positions, but densities up to about 1,000 spots per membrane have been achieved. The grid can be freely defined. Up to 44 amino acid derivatives can be used, which allows for the use of natural and non-natural amino acids in one run.

As automation of the SPOT synthesis procedure with the spot robot ASP 222 enables the handling of large arrays, powerful software is necessary to generate the sequences. Sequences can be entered in the software itself or imported from text files in one-letter code from other sources. There are several modes for generating peptides from parent protein or peptide sequences:

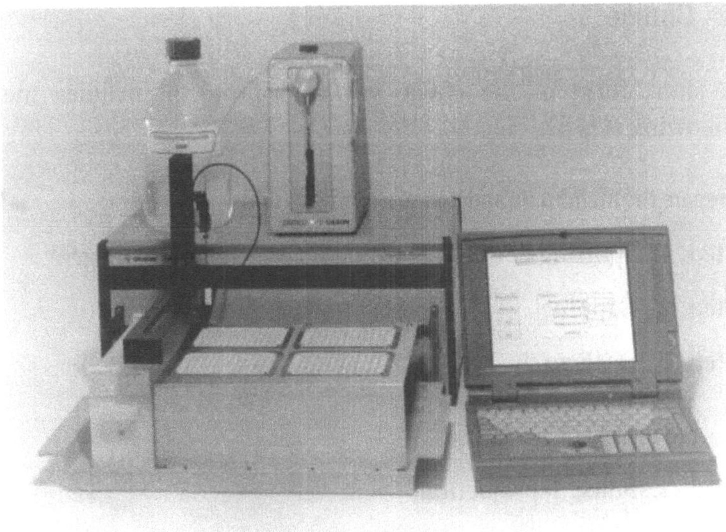

Fig. 1. The ASP 222 automated spot robot

1. Individual peptides (entered one by one)
2. Analogue sequences (by defined amino acid replacement)
3. Overlapping peptides (generated from a protein sequence)
4. Overlapping peptides of different length (from a protein sequence)
5. Peptide libraries with a maximum of two mixed positions (libraries)

Sequences are entered as one-letter code (OLC). The instrument can handle up to 44 amino acid derivatives defined as uppercase OLC plus B, O, X, Z, and lowercase OLC except b, o, x, z. This allows the use of natural and unnatural amino acids in one synthesis run.

The maximum length allowed by the software is 40 residues, but practically the sequences should not be longer than 15–20 residues.

Outline

A short form of the entire synthesis protocol includes the following steps:

Prepare the membrane and define a synthesis grid

Step 1. Spotting of the first amino acid to generate the grid

Step 2. Capping

Step 3. DMF wash

Start the synthesis cycles

Step 4. Fmoc deprotection

Step 5. DMF wash

Step 6. Ethanol wash

Step 7. Air drying of the membrane

Step 8. Spotting of activated amino acids and waiting for reaction time

Step 9. Capping, optional

Step 10. DMF wash

Repeat synthesis steps 4–10 until the desired peptides have been assembled.

Final Fmoc deprotection

At the end of the synthesis, the peptides carry the Fmoc-protection group, which is removed using synthesis steps 4–7.

Side-chain deprotection

Side-chain protection groups must be removed in a different protocol using a mixture of trifluoroacetic acid (TFA) in dichloromethane (DCM) and appropriate scavengers.

After the final washing of the membrane, the covalently bound peptides are ready to be screened for biological activity.

Materials

Equipment

- Auto-Spot Robot ASP 222, (ABIMED Analysen-Technik GmbH, Langenfeld, Germany or, outside of Germany and Austria: INTAVIS AG, Bergisch Gladbach, Germany)

Synthesis membranes

- Membranes for automated spot synthesis, 130×90 mm, derivatized purified cellulose. The membranes are now derivatized with a polyethylene glycol spacer linked via a base-stable bond, up to 400 nmol/cm^2. The spacer has a length of 8–10 ethylene glycol units. There are clear benefits with this spacer:
- Stability: aqueous pH 0 to pH 14 at ambient temperature
- Greater distance to cellulose carrier
- Hydrophilic spacer
- Reduced background problems
- No initial Fmoc deprotection is required (ABIMED Analysen-Technik GmbH, Langenfeld, Germany or, outside of Germany and Austria: INTAVIS AG, Bergisch Gladbach, Germany)
- Wick filters, 40/pack

Chemicals

Chemicals for about 4 membranes, 4×96 or 4×384 10-mer peptides:

Dimethylformamide (DMF, lab or synthesis grade)	5 l
Piperidine p.a.	250 ml
Ethanol (ETOH) p.a.	1 l
Acetic anhydride	100 ml
Trifluoroacetic acid (TFA)	100 ml
Dichloromethane (DCM)	1 l
Triisopropylsilane	5 ml
Hydroxybenzotriazole (HOBt)	10 g
1-Methyl-2-pyrrolidone (1-methyl-2-pyrrolidone, dry, puriss. p. a. quality)	250 ml
Diisopropyl carbodiimide (DIC)	25 ml
Molecular sieve 4Å	250 g
Bromophenol blue (BPB)	1 g

Source: Fluka, Merck or Sigma

Amino acid derivatives

ABIMED and INTAVIS supply convenient cartridges of amino acids (0.5 mmol), each in a dry and not activated form, for use in solid-phase synthesis with the ASP 222. Stock solutions in high quality 1-methyl-2-pyrrolidone with HOBt added are stable at 4 °C for about a week. For synthesis, aliquots of these stock solutions must be activated by addition of DIC. Activated derivatives are stable for about a day. More details about the amino acid derivatives are given in Chapter 2.

Derivative	MW	0.5 mmol
Fmoc-Ala-OH	311.3	155.7 mg
Fmoc-Arg(Pmc)-OH	662.8	331.4 mg
Fmoc-Asn(Trt)-OH	596.7	298.4 mg
Fmoc-Asp(OtBu)-OH	411.5	205.8 mg
Fmoc-Cys(StBu)-OH	431.6	215.8 mg
(Fmoc-Cys(Trt)-OH	585.7	292.9 mg
Fmoc-Gln(Trt)-OH	610.7	305.4 mg
Fmoc-Glu(OtBu)-OH	425.5	212.8 mg
Fmoc-Gly-OH	297.3	148.7 mg
Fmoc-His(Trt)-OH	619.7	309.9 mg
Fmoc-Ile-OH	353.4	176.7 mg
Fmoc-Leu-OH	353.4	176.7 mg
Fmoc-Lys(Boc)-OH	468.5	234.3 mg
Fmoc-Met-OH	371.5	185.8 mg
Fmoc-Phe-OH	387.4	193.7 mg
Fmoc-Pro-OH	337.4	168.7 mg
Fmoc-Ser(tBu)-OH	383.4	191.7 mg
Fmoc-Thr(tBu)-OH	397.5	198.8 mg
Fmoc-Trp(Boc)-OH	526.6	263.3 mg
Fmoc-Tyr(tBu)-OH	459.6	229.8 mg
Fmoc-Val-OH	339.4	169.7 mg
Fmoc-β-Ala	311.3	155.7 mg

Use of other pre-activated amino acids

Instead of HOBt-esters activated in situ with DIC you can also use pre-activated amino acids such as pentafluorophenyl esters (Opfp). These are somewhat less reactive and more expensive than HOBt esters generated in situ.

Other reagents and wash solutions

- Capping reagent:
  2 % Acetic anhydride in dry DMF
  Add to 150 ml of dry DMF 3 ml acetic anhydride.
- Deprotection of the Fmoc group:
  20 % Piperidine in dry DMF (15 ml per wash cycle)
- Ethanol:
  Final wash solution before drying, 15 ml per wash cycle.
- Reagent for final side-chain deprotection (for 10 ml):
  5 ml Trifluoroacetic acid (TFA)
  5 ml Dichloromethane (DCM)
  300 µl Triisobutyl silane
  200 µl Water
- Bromophenol blue solution:
  100 ml DMF
  1 g (w/v) Bromophenol blue (BPB)

Procedure

Warning!

Hazardous chemicals. During handling of chemicals and solvents, a lab coat, gloves and eye protection must be worn. Please observe the safety regulations concerning chemicals used for the synthesis.

Preparation of amino acid derivative stock solutions

1. To each cartridge (or 0.5 mmol amino acid derivative) add 1.0 ml of a solution of 0.75 mmol/ml HOBt in dry 1-methyl-2-pyrrolidone (for 20 cartridges you will need 2.3 g HOBt in 20 ml 1-methyl-2-pyrrolidone).
2. Shake well to dissolve amino acid derivatives.
3. Add 1-methyl-2-pyrrolidone to each cartridge to achieve a total volume of approximately 1.5 ml.
4. Dissolve the amino acids by vigorous shaking if not yet dissolved.
5. Aliquot 450 µl of the amino acid stock solution into labeled reaction tubes of 2 ml (Eppendorf tubes), three tubes for each amino acid.

6. Store the solutions at 4 °C.

7. You have now prepared 0.33 M stock solutions of non-activated derivatives; these can be kept at 4 °C for about 1 week. For longer storage, use a –20 °C freezer.

Activation of amino acid derivatives

8. Let aliquots of the derivatives warm up to room temperature, one tube for each amino acid used during the whole day. **Exception:** As activated Fmoc-Arg(Pmc) decomposes rapidly, this derivative should be freshly prepared just prior to each synthesis cycle.

9. Prepare an activator stock solution of 1.1 mmol/ml DIC in 1-methyl-2-pyrrolidone (Example: add 0.8 ml DIC to 4.2 ml 1-methyl-2-pyrrolidone). This solution is stable for 1 day.

10. Activate a 450-µl amino acid stock solution with 150 µl activator stock solution (DIC in 1-methyl-2-pyrrolidone). The 600-µl activated amino acid solution has a final concentration of 0.25 mmol/ ml.

11. Shake the mixture and wait for 10–30 min to complete activation. **Note:** During activation, urea crystals may form within 10–20 min. To remove the crystals from the solution, it is advisable to spin the tubes in a centrifuge and transfer supernatant to a fresh tube.

12. Place the tubes in the amino acid rack of the Auto-Spot Robot ASP 222. See the specific section in the Auto-Spot manual for more details.

Preparation of the membrane and definition of a synthesis grid

The spot method is used to synthesize peptides at defined positions of a homogeneous membrane. This requires loading of the membrane with anchor groups at the respective positions. The membranes come derivatized with a polyethylene-glycol spacer linked via a base-stable bond. As the spacer is already a free amino function, synthesis can be started without prior Fmoc deprotection. The grid should still be generated by spotting β-alanine or the C-terminal amino acid of each peptide at a volume about 30 % below the volume used for synthesis. The volume spotted throughout the synthesis should be higher than in the first cycle to avoid incomplete couplings at the border of spots during synthesis.

Preparation of the membrane

1. Mount each synthesis membrane on top of a wick-filter on the holding plate. **Note:** The thick wick-filter is used to absorb excess liquid. Replace it after each synthesis cycle to avoid contamination. If you decide to work without wick-filters, you must clean the holder with alcohol in each cycle. Place one sheet of the synthesis membrane on each wick-filter. When working with more than one membrane, you must mark them with a graphite pencil to ensure unambiguous positioning in each synthesis cycle. Any other ink is dissolved during the wash cycle.

Defining the grid

2. Start the system (Auto-Spot Robot ASP 222) and start the grid definition cycle. The instrument will now deliver Fmoc-β-alanine (or the C-terminal amino acids) to each spot on the membranes to define the grid. Repeat this step at least one time. **Note:** The volume delivered will determine the spot size, which is limited by the grid distance selected. Use 0.6 μl for an 8×12 grid and 0.2 μl for a 16×24 grid.

Capping

3. Wait 30 min or until all spots are dry.
4. Take the membranes off the holder and wash each one separately in a solution of 2 % acetic anhydride in DMF for capping of the remaining area between spots. Immerse the membranes for 30 s with capping solution. Decant and repeat for 2–5 min with fresh solution (15 ml capping solution for each wash).

DMF wash

5. Wash the membranes with 15 ml DMF for 1 min once and 3–4 times for 2 min each.

Start of the synthesis cycles

The following procedures listed are carried out for each synthesis cycle until the largest peptide of the series has been assembled.

Fmoc deprotection

1. Immerse the synthesis membranes two times (5–10 min) with 20 % piperidine in DMF for removal of the Fmoc-protection group (15 ml solution for each wash).

DMF wash

2. Wash the membranes with 15 ml DMF for 1 min once and three times for 2 min. **Note:** If you want to visualize the spots, wash the membranes with 150 µl bromophenol blue in 15 ml DMF. This is also a good test for complete removal of piperidine, as the spots will not stain with piperidine present.

Ethanol wash

3. Wash the membranes with 15 ml ethanol two times for 2 min each.

Air drying

4. Press the membranes between several layers of clean filter paper to remove excess liquid. Then dry the membranes with a stream of cold air.
5. Mount membranes on the holder, let the ASP spot the activated amino acids and wait for a reaction time. The standard reaction time is of the order of up to 20 min. Repeat this step at least once (double coupling) to increase the coupling yield and thus the synthesis quality.
6. (Optional step!) A capping step after each synthesis cycle is not mandatory. Note that the capping step will acetylate some peptides if they are not all of the same length. In this case, a capping step should also be performed after final Fmoc deprotection, but before side-chain deprotection.
7. Wash the membranes two times for 1 min with 15 ml DMF.
8. Repeat the synthesis cycle (steps 1–7 "Start of the Synthesis Cycles") until the desired peptides have been assembled. The peptides are now fully assembled and carry either the Fmoc-protection group or an acetyl group from capping.

Final Fmoc deprotection

1. Immerse the synthesis membranes two times (5–10 min) in 20 % piperidine in DMF for deprotection of the Fmoc-protection group (15 ml solution for each wash).
2. Wash the membranes with 15 ml DMF for 1 min once and three times for 2 min.

3. Wash the membranes with 15 ml ethyl alcohol two times for 2 min each.

4. Press the membranes between several layers of clean filter paper to remove excess liquid. Then dry the membranes with a stream of cold air. **Note:** If your assay requires acetylated peptides, you can include another capping reaction after Fmoc removal.

Side-chain deprotection

Warning: Hazardous chemicals!

TFA is a strong acid. It can cause severe burns, is corrosive and vapors are harmful if inhaled. Mixing with DMF is a strongly exothermic reaction and can cause splashing. You must observe the following safety guidelines:

- Always work in a fume hood,
- Wear lab coat, gloves and eye protection,
- Never mix TFA and DMF.
- Do not breathe TFA vapor.
- Observe safety regulations applicable to your laboratory.

Side-chain protection groups must be removed in a different procedure using a mixture of TFA in dichloromethane DCM and appropriate scavengers. The reaction is carried out in a polypropylene box with a lid.

The membranes must be thoroughly dried before the peptides can be side-chain deprotected. Let the membranes dry overnight or use a vacuum desiccator.

1. Prepare 10 ml of reagent for final side-chain deprotection.

2. Let the membranes react with this solution for about 1 h.

3. Wash four times with 20 ml DCM.

4. Wash four times for 2 min with 15 ml DMF.

5. Wash two times for 2 min with 15 ml ethanol.

6. Dry the membranes with cold air.

Results

The synthesis is now completed and the membranes with covalently bound peptides are ready to screen for biological activity. The membranes should be stored dry and cold.

Troubleshooting

- Solvent preparation
  The solvent used should be of high quality. Especially 1-methyl-2-pyrrolidone for the solution of the amino acid derivatives must be dry and free of amines.
  Buy 1-methyl-2-pyrrolidone stored over molecular sieve or add molecular sieve corresponding to about 5 % of the solvent volume. Let it stand for a couple of days.
  DMF for washing does not require any special treatment if the solvent is of synthesis grade for peptides.

- Spot size (amino acid volume)
  To avoid border effects, the spots positioned during generation of the grid should be smaller than the areas soaked during synthesis. This is accomplished by using a smaller spot volume during generation of the grid than during synthesis.

  Recommended volumes during grid definition cycle:
  - 8×12 grid: 0.6 μl
  - 16×24 grid: 0.2 μl

  Recommended volumes during synthesis:
  - 8×12 grid: 0.8 – 1.0 μl
  - 16×24 grid: 0.3 μl

- Reaction time
  Reaction times between repeats are defined in the main menu before synthesis start.
  Reaction time of about 30 min should be selected when synthesizing less than 100 spots. Fore more than 100 spots, 20 min are sufficient, for more than 600 spots no reaction time needs to be specified between repeats.

- Number of repeats
  Small volumes, such as those used for small spots, evaporate quickly. To ensure complete coupling, we recommend distributing reagents two or even tree times for one synthesis cycle.

  Bromophenol blue (BPB) monitoring
  Free amino groups can be visualized with the dye BPB.
  The dye binds to amines as a deep blue anion and provides an elegant method to visualize coupling reactions.
  Prepare 1–2 ml of a 1 % stock solution of BPB in DMF.
  Add 10 µl of the BPB stock to 1 ml of 1-methyl-2-pyrrolidone to check for amines. If the solution stays yellow or light green, the 1-methyl-2-pyrrolidone can be used.
  During synthesis, add 150 µl of the BPB stock to the last DMF wash after Fmoc deprotection.
  The color of the wash solution should be yellow or light green.
  If the solution turns blue, repeat the DMF washing. Only if the solution stays yellow or light green is the dye bound by amino groups in the membrane. The blue spots indicate free amino groups. The membrane is now prepared for the next coupling cycle. During synthesis, the fresh, activated amino acid will displace the blue dye by coupling to the amino groups. Blue or green blue spots show incomplete coupling. Nevertheless, the actual coupling yield cannot be determined. After coupling, the spots are never colorless. Efficient coupling is indicated by a color spectrum between yellow and green.
  The dye stays on the spots during the ethanol wash and drying. For documentation of the synthesis, you can photocopy the membranes after a few synthesis cycles, especially if there are doubts about some of the peptides.

References

Frank R (1992) Spot synthesis: an easy technique for the positionally addressable, parallel chemical synthesis on a membrane support. Tetrahedron 48:9217–9232

Suppliers

ABIMED Analysen-Technik GmbH
(e-mail: info@abidmed.de, www.abimed.de,
Tel.: +49 (0)2173-8905-0, Fax: +49-2173-8905-77)
Raiffeisenstraße 3, 40764 Langenfeld, Germany

INTAVIS Bioanalytical Instruments AG
(e-mail: info@intavis.com, www.intavis.com,
Tel.: +49-2204-84 32 50, Fax: +49-2204-84 32 58)
Friedrich-Ebert-Strasse, 51429 Bergisch Gladbach, Germany

Epitope Mapping of Antibodies with Solid-Phase Oligopeptides

JOACHIM KOCH, MICHAEL MAHLER, and MARTIN BLÜTHNER

Introduction

Epitope mapping represents a powerful tool not only for basic research but also for therapeutic applications. The knowledge of the exact binding site of antibodies on their antigens could ease the development of new vaccination strategies against pathogens. Furthermore, these insights could be used for the development of new therapeutic approaches to remove pathogenic autoantibodies from the serum of autoimmune patients.

The majority of mammalian species express five different classes of antibodies (immunoglobulins): IgG, IgM, IgA, IgD, and IgE. Most of the antibodies used for epitope mapping studies belong to the IgG class since they are most abundant in serum and are responsible for the neutralization of microbes and toxins. Furthermore, most autoantibodies also belong to the latter type. IgG is a monomer comprised of two γ heavy chains and two light chains leading to a molecular weight of the antibody of 150 kDa. An IgG antibody has two antigen binding sites. In serum, it makes up about 80 % of the total serum immunoglobulins. Antibodies of other immunoglobulin classes could also be used for epitope mapping studies, but it might be necessary to adjust the assay

J. Koch (✉) (e-mail: joachim.koch@em.uni-frankfurt.de,
Tel.: +49-69-79829273, Fax: +49-69-79829495)
Forschungsstelle Hantaviren, Heidelberger Akademie der Wissenschaften
Current Address: Institut für Biochemie, Biozentrum N210/20,
Marie-Curie-Straße 9, 60439 Frankfurt am Main, Germany
M. Mahler, Institut für Molekulare Genetik, Universität Heidelberg
Current Address: Pharmacia Deutschland GmbH, Munzinger Straße 7,
79111 Freiburg, Germany
M. Blüthner
Institut für Molekulare Genetik, Universität Heidelberg, Germany

Springer Lab Manual
J. Koch, M. Mahler (Eds.) Peptide Arrays on Membrane Supports

conditions since all antibody classes differ in size, number of antigen binding sites per molecule, stability and serum titer.

Oligopeptides synthesized on cellulose membranes according to the SPOT method are useful for the identification and characterization of linear epitopes recognized by monoclonal or serum antibodies (Frank 1992; Korth et al. 1997; Kramer et al. 1997; Kneissel et al. 1999; Mahler et al. 2000). The synthesis of oligopeptides, usually 5–20 amino acids in length, can be performed on activated cellulose membranes either manually (up to 100 peptides; see Chap. 3) or using a pipetting robot (several thousand peptides; see Chap. 4). When using a pipetting robot, it is possible to cover complete amino acid sequences of large proteins in the form of overlapping "protein fragments." Following completion of the synthesis, the peptide arrays can be probed with monoclonal or serum antibodies allowing for determination of the reactive peptides. Finally, this approach leads to the identification of the corresponding binding sites within the protein. In case of polyclonal serum samples, it might be necessary to distinguish between adjacent and overlapping epitopes recognized by antibodies with different specificities (Mahler et al. 2000). This can be achieved by synthesizing and probing overlapping peptides of decreasing lengths covering the area of interest.

After identification of the epitope sequence(s), all amino acids within the epitope can be successively replaced by alanine (alanine-walking) or glycine (glycine-walking). Alternatively, replacement of every amino acid position within the epitope region with any of all 20 amino acids can be performed to further characterize the binding site (see Chap. 9; Liu et al. 1999; Blüthner et al. 2000). In rare cases side-chain modification of certain amino acids within an epitope or the use of unusual amino acids is required to increase the binding affinities of antibodies (see Chap. 10). In many epitope mapping studies it has been shown that the character of epitopes is quite diverse.

- Some antibodies recognize linear epitopes of less than 10 amino acids located in close proximity to each other. These epitopes can easily be identified using oligopeptides since all amino acids contributing to the interaction are located within one peptide (Korth et al. 1997).

- Furthermore, conformational epitopes located in rather simple 3D structures like α-helices can be identified and further characterized by mutational analysis of identified binding sites (see Chap. 9; Kneissel et al. 1999; Liu et al. 1999; Blüthner et al. 2000).
- However, when the amino acids comprising the epitope are distributed over distant parts of the primary sequence of the protein, thereby forming an epitope by tertiary structure formation, those discontinuous or conformational epitopes are accessible by the SPOT method only when antibody binding to partial epitopes shows very high affinity (Reinecke et al. 1999).

Although it has been shown that this method is fast, convenient, and reliable, it should not be the initial strategy for a protein-antibody interaction study. Standard immunodetection procedures like ELISA, immunoblotting, deletion mutant studies or the use of gene fragment phage-display libraries should be used for the initial localization of the interaction domains on the protein (Blüthner et al. 1996; Fack et al. 1997).

Outline

The basic procedure for the identification of epitopes within a certain protein sequence can be divided into several steps which are outlined in Fig. 1. Following the detection of immunoreactive areas on the protein by standard detection methods, 15-mer peptides, offset by three to five amino acids, can be probed with a monoclonal antibody or polyclonal serum analogous to the procedure used for immunoblots (see Chaps. 3 and 4). The reactive peptides of the epitope region are then synthesized as 15-mer peptides offset by one amino acid and probed again with the monoclonal antibody or polyclonal serum (see Chaps. 3 and 4). When serum antibodies are used for epitope mapping, the length of the overlapping peptides can be reduced to 12-, 10-, 8-, 7-, 6-, or 5-mers to distinguish between adjacent or overlapping epitopes of the different antibody specificities.

More detailed information about the character of the epitope can be obtained from mutational analyses or theoretical structure predictions (see Chap. 9).

Step 1
Binding studies using standard biochemical and molecular biological methods, to identify the subdomain of the antigen containing the binding site.

↓

Step 2
Optional:
Rough scan of the target protein (or subdomain) by probing overlapping 15mer peptides offset by 3 to 5 amino acids.

↓

Step 3
Fine scan of the target protein or the reactive peptide stretches identified in the rough scan with overlapping 15mer peptides offset by one amino acid.

↓

Step 4
Synthesis and probing of shorter peptides (down to 5mers) based on sequences being reactive in the fine scan to distinguish between overlapping epitopes.

↓

Step 5
Subsequent exchange of every single amino acid position within the sequence against alanine (alanine walking) or glycine (glycine walking).

↓

Step 6
Subsequent exchange of every single amino acid position within the epitope against all other naturally occurring 20 amino acids.

Fig. 1. Outline of the standard protocol for an epitope mapping study using solid-phase oligopeptides. Steps 5 and 6 are discussed in more detail in Chapter 9

Materials

Buffers

- Tris-buffered saline (TBS, pH 7.6):
  10 mM Tris/HCl
  150 mM NaCl
- TBS, 0.2 % Tween 20 (TBS-T):
  TBS (pH 7.6)
  0.2 % (v/v) Tween 20

- Antibody dilution buffer:
  TBS (pH 7.6)
  10 % (v/v) 10× Blocking buffer concentrate
  0.05 % (v/v) Tween 20
  3 % (v/v) Inactivated horse-donor-serum
  5 % (w/v) Sucrose
- Blocking solution A:
  TBS (pH 7.6)
  3 % (w/v) BSA
- Blocking solution B:
  TBS (pH 7.6)
  2 % (w/v) Skim milk powder
- Blocking solution C:
  TBS (pH 7.6)
  2 % (w/v) Skim milk powder
  0.2 % (v/v) Tween 20
- Blocking solution D:
  TBS (pH 7.6)
  50 % (v/v) Inactivated horse-donor-serum
  10 % (v/v) 10× Blocking buffer concentrate
  0.2 % (v/v) Tween 20
  150 mM sucrose
- Blocking buffer concentrate (SU-07–250, 10×; Genosys, Cambridge, UK)
- Other blocking systems, see Chap. 11
- Regeneration buffer A:
  8 M urea
  1 % (w/v) Sodium dodecyl sulfate (SDS)
  0.1 % (v/v) β-mercaptoethanol
- Regeneration buffer B:
  50 % (v/v) Ethanol
  10 % (v/v) Acetic acid in distilled H_2O

Solutions

- β-mercaptoethanol
- ECL Western blotting detection reagents (Amersham Pharmacia Biotech, Piscataway, NJ, USA)

Note: Self-made ECL detection reagents work equally well.

- Solution A:
  200 µl luminol (250 mM)
  88 µl *p*-coumaric acid (90 mM)
  2 ml Tris/HCl (1 M, pH 8.5)
  17.7 ml ddH_2O
- Solution B:
  12 µl H_2O_2 (30 %)
  2 ml Tris/HCl (1 M, pH 8.5)
  18 ml ddH_2O
  Mix equal volumes of solution A and B prior to use.
- *N,N*-dimethylformamide (DMF; p.a.)
- Tween 20

Appropriate horseradish peroxidase (HRP)- or alkaline phosphatase (AP)-conjugated secondary antibody

- Amersham Lifescience Hyperprocessor (Amersham Pharmacia Biotech)
- Hair dryer
- Hyperfilm (ECL; Amersham Pharmacia Biotech)
- Hypercassette (Amersham Pharmacia Biotech)
- Nescofilm (e.g. Nippon Shoji Kaisha Ltd., Osaka, Japan)
- Side-to-side shaker

Procedure

Epitope mapping

After completion of peptide synthesis (see Chaps. 3 and 4), the peptide spots are absolutely water free and thus have to be prepared for the epitope-mapping procedure.

1. Place the membrane in 20 ml 100 % ethanol. Rehydrate the membrane by adding 150 ml TBS to the membranes over a period of 30 min. **Note:** Hydrophobic peptides may need more time to hydrate. Make sure that the white color of all spots has completely disappeared, indicating a successful rehydration.

2. Block unspecific binding sites with blocking buffer for 2 h at room temperature or overnight at 4 °C on a side-to-side shaker (see Fig. 2, step 1). Different blocking conditions can

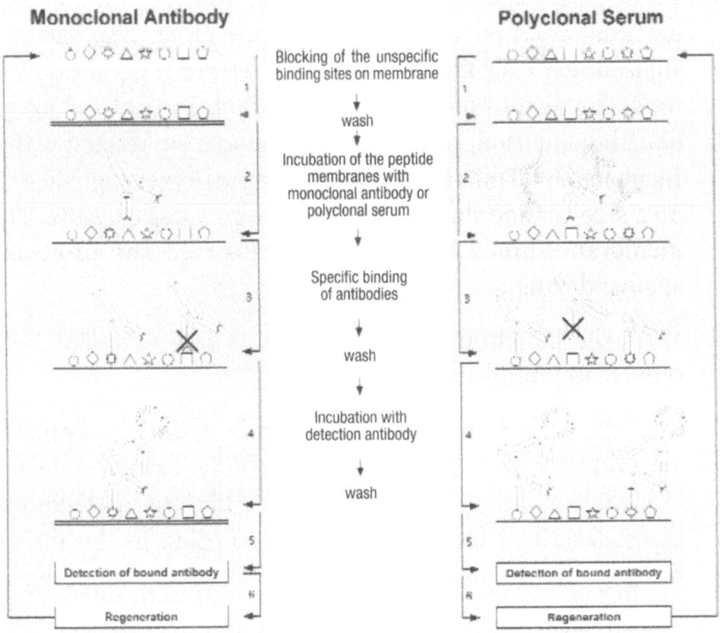

Fig. 2. Procedure of an epitope mapping study with monoclonal antibodies (*left*) or polyclonal serum (*right*) using overlapping solid-phase oligopeptides of the target protein

be compared to obtain an optimal signal-to-noise ratio. The following blocking solutions, increasing in "stringency," are recommended.

a) Blocking solution A: 3 % (w/v) BSA in TBS
b) Blocking solution B: 2 % (w/v) skim milk powder in TBS
c) Blocking solution C: TBS with 2 % (w/v) skim milk powder, 0.2 % (v/v) Tween 20
d) Blocking solution D: TBS with 50 % (v/v) horse serum, 10 % (v/v) 10× blocking buffer concentrate, 0.2 % (v/v) Tween 20, 150 mM sucrose

Note: Based on our experience, blocking solution C works best for most monoclonal antibodies, blocking solution D for serum antibodies or high-affinity monoclonal antibodies.

3. Wash the membrane once for 10 min with TBS/0.2 % (v/v) Tween 20 (TBS-T).

4. Dilute antibody of interest in antibody dilution buffer in case blocking solution D was used. In any other case, use the solution used for blocking (see step 2). We suggest diluting monoclonal antibodies to a final concentration of 2–5 µg/ml blocking solution, polyclonal sera should be diluted 1:100. Incubate for 90 min at room temperature or overnight at 4 °C on a side-to-side shaker (see Fig. 2, steps 2 and 3). **Note:** The membrane should be completely immersed and protected against drying.
5. Wash the membrane three times for 5 min with TBS-T to remove unbound antibodies.
6. Incubate the membrane for 75 min with a secondary antibody diluted in antibody dilution buffer or in the selected blocking solution (see step 4) using conditions analogous to those described for immunoblots according to the manufacturer's instructions (see Fig. 2, step 4). **Note:** Depending on the detection system, HRP- as well as AP-conjugated antibodies can be used for detection. Based on our experience the use of HRP-conjugated detection antibodies is most convenient.
7. Wash the membranes three times for 5 min with TBS-T followed by three washes for 5 min with TBS.
8. Remove excessive buffer from the membranes by placing white tissue paper on the membrane. **Note:** Do not use recycled tissue since it might interfere with the detection system. To avoid damage to the peptide coats, do not wipe or press tissue onto the membrane. Avoid drying of the membrane.
9. Detect the spots with chemiluminescence (see Fig. 2, step 5).

Detection of reactive peptide spots

Since a minimum of two different incubations, sample plus negative control, are recommended for every analysis, the ECL detection system is the preferred visualization agent. When using the AP system, precipitated color might not be removed from the membrane by regeneration. Furthermore, the ECL system has a high sensitivity and allows for convenient adjustment of the appropriate exposure time to obtain a maximum signal-to-noise ratio.

1. Place the membrane onto a Nescofilm larger in size than the membrane and place both into a Hypercassette.
2. Mix ECL detection reagents according to the manufacturer's instructions, apply the mixed solution onto the membrane, and incubate for 1 min at room temperature. Since the Nescofilm has a hydrophobic surface, the liquid will stay on the membrane. This will minimize the required volume of detection reagent to immerse the membrane. Remove excessive ECL solution with white tissue paper (see "Epitope mapping," step 8). **Note:** When self-made ECL detection reagents are used, mix equal volumes of freshly prepared solution A and B prior to use as described in the Materials section and incubate the membrane for 1 min at room temperature.
3. Cover the membrane on the Nescofilm with a transparency (such as used for over-head presentations), apply an ECL-film, and expose at room temperature. The reactive peptide spots can be matched to their corresponding overlapping

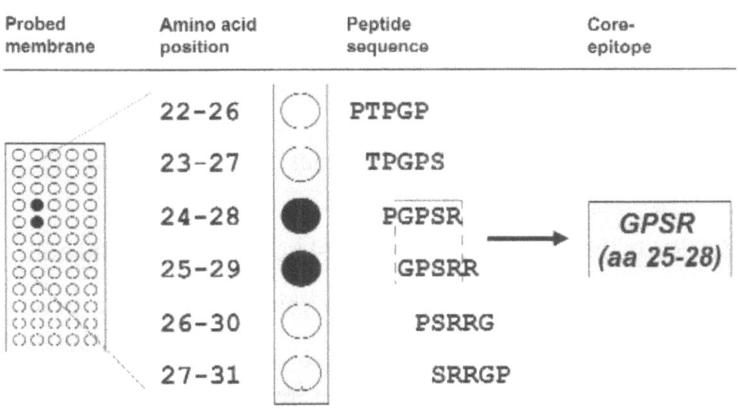

Fig. 3. Identification of a core epitope on centromere-protein A. Single immunoreactive peptide spots can be assigned to their corresponding amino acid sequences. The core epitope (GPSR; amino acid 25–28) can be identified by matching of the overlapping peptides. The minimal matching sequence stretch contained in all immunoreactive peptides represents the core epitope

sequences. An example of determination of the amino acids essential for antibody binding is given in Fig. 3. **Note:** Start with an exposure time of 1 min to get an impression of the signal intensity. Try other exposure times until a maximum signal-to-noise ratio is obtained.

4. Transfer the membrane into distilled H_2O and store at 4 °C until it is subjected to regeneration (see "Regeneration of the membrane" and Fig. 2, step 6). **Note:** The membrane can be stored in water for several days prior to regeneration.

Regeneration of the membrane

The regeneration procedure of the membrane is mainly based on two principles. Regeneration buffer A contains SDS and β-mercaptoethanol, both leading to denaturation of the peptide-bound antibodies, and regeneration buffer B, due to its low pH (2–3), leads to elution of the antibodies. All steps of the regeneration procedure are carried out at room temperature.

1. Rinse the membrane three times for 5 min with 50 ml distilled water.
2. Wash the membrane three times for 5 min with 25 ml DMF.
3. Rinse the membrane three times for 5 min with 50 ml distilled water.
4. Wash the membrane three times for 10 min with 25 ml regeneration buffer A.
5. Wash the membrane three times for 10 min with 25 ml regeneration buffer B.
6. Rinse the membrane twice for 10 min with 25 ml ethanol.
7. Dry the membrane with cold air using a regular hair dryer.

The membrane can either be stored at –20 °C in a sealed plastic bag or it can directly be used for another epitope mapping study (see "Epitope mapping"). **Note:** The membranes can usually be regenerated up to ten times without loss of signal intensity. To check for successful regeneration, incubate the regenerated membrane directly with the ECL detection reagents and expose an ECL film (see "Detection of reactive peptide spots"). No signal should be obtained after 30 min of exposure. After three washes in 50 ml TBS, the membrane can directly be used for another epitope mapping experiment (proceed directly with the "Epitope mapping," step 2).

Results

The CENP-A model system

Centromere protein A (CENP-A) is a 17-kDa protein consisting of 140 amino acids. It has been shown that the C-terminal part of CENP-A protein shares approximately 70 % amino acid sequence homology to the histone protein H3 (Sullivan et al. 1994). Patients suffering from the limited form of systemic sclerosis (lSSc) frequently develop autoantibodies against the centromere proteins, including CENP-A (Moroi et al. 1980; Tan et al. 1980; Guldner et al. 1984; Earnshaw et al. 1985). Using several biochemical and molecular biological methods, it has been shown that the B-cell autoimmune response is directed against epitopes

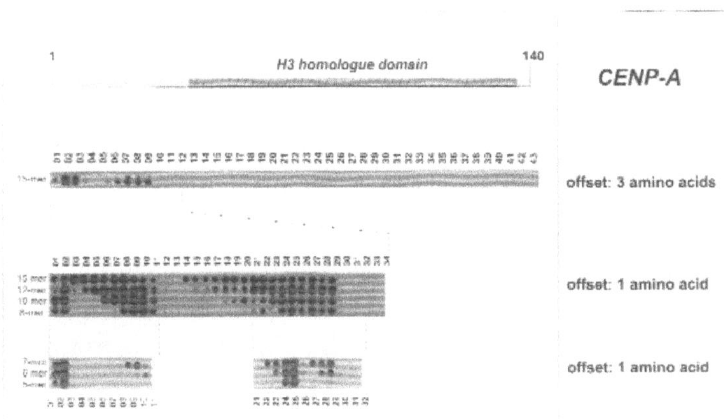

Fig. 4. Epitope-mapping of a human autoimmune serum using solid-phase oligopeptides. For the first scan, 15-mer solid-phase oligopeptides of the N-terminus of CENP-A offset by three amino acids were synthesized on a cellulose membrane and probed with the serum (*top*). To discriminate between the different antibody specificities, shorter peptides were subjected to analysis (*middle* and *bottom*)

of the N-terminal domain of the histone-like protein (Muro et al. 1996, 2000). To elucidate the precise binding sites of the different antibody specificities of lSSc patient sera, 15-mer peptides offset by three amino acids representing the entire sequence of CENP-A were synthesized and probed with the serum antibodies (Fig. 4). We were able to show that the immune response of these lSSc patients is exclusively directed against a stretch of 45 amino acids at the N-terminus of CENP-A (Mahler et al. 2000). Since these autoimmune sera frequently contain a variety self-reactive antibodies against adjacent or overlapping epitopes, it was necessary to analyze the binding to shorter peptides of the immunoreactive area. For this approach the N-terminal region of CENP-A was synthesized as 15-, 12-, 10- and 8-mer peptides offset by only one amino acid and probed again with the patient sera. Surprisingly, it was not possible in all cases to discriminate between the binding sites of all serum antibodies. Therefore, 7-, 6- and 5-mer oligopeptides were subjected to analysis. By this second approach we were able to identify the exact binding sites of the serum antibodies of 19 anti-centromere sera (Mahler et al. 2000). An example of the epitope-mapping procedure is illustrated in Fig. 3.

Troubleshooting

- No signal is obtained after 30 min of exposure:
  Check the detection system and use less stringent blocking conditions (see "Epitope mapping," step 2). If no binding occurs, this may indicate a discontinuous binding site or very low affinity binding.

- Unspecific signals and high signal-to-noise ratio:
  Increase the stringency of the blocking conditions (see "Epitope mapping," step 2) and make sure that the antibodies used are of high purity. Increase the dilution factor of the antibodies.

- Antibodies bound to the peptide spots cannot be removed by regeneration:
  This phenomenon has been observed only in very rare cases with very high affinity antibodies. Synthesized membranes can only be used for one incubation with the antibody to be tested.

Acknowledgements. The experiments were supported by grant BL 483/1-1 to M. Büthner from the Deutsche Forschungsgemeinschaft. We thank Prof. Dr. E.K.F. Bautz for continued advice and valuable suggestions.

References

Blüthner M, Bautz EKF, Bautz FA (1996) Mapping of epitopes recognized by PM/Scl autoantibodies with gene-fragment phage display libraries. J Immunol Methods 198:187–198

Blüthner M, Mahler M, Müller DB, Dünzl H, Bautz FA (2000) Identification of an alpha-helical epitope region on the PM/Scl-100 autoantigen with structural homology to a region on the heterochromatin p25beta autoantigen using immobilized overlapping synthetic peptides. J Mol Med 78:47–54

Earnshaw WC, Rothfield N (1985) Identification of a family of human centromere proteins using autoimmune sera from patients with scleroderma. Chromosoma 91:313–321

Fack F, Hügle-Dörr B, Song D, Queitsch I, Petersen G, Bautz EKF (1997) Epitope mapping by phage display: random versus gene-fragment libraries. J Immunol Methods 206:43–52

Frank R (1992) SPOT synthesis: an easy technique for the positionally addressable, parallel chemical synthesis on a membrane support. Tetrahedron 48:9217–9232

Guldner HH, Lakomek HJ, Bautz FA (1984) Human anti-centromere sera recognise a 19.5 kD non-histone chromosomal protein from HeLa cells. Clin Exp Immunol 58:13–20

Kneissel S, Queitsch I, Petersen G, Behrsing O, Micheel B, Dübel S (1999) Epitope structures recognised by antibodies against the major coat protein (g8p) of filamentous bacteriophage fd (Inoviridae). J Mol Biol 288:21–28

Korth C, Stierli B, Streit P, Moser M, Schaller O, Fischer R, Schulz-Schaeffer W, Kretzschmar H, Raeber A, Braun Ehrensperger F, Hornemann S, Glockshuber R, Riek R, Billeter M, Wuthrich K, Oesch B (1997) Prion (PrPSc)-specific epitope defined by a monoclonal antibody. Nature 390:74–77

Kramer A, Keitel T, Höhne W, Schneider-Mergener J (1997) Molecular basis of binding promiscuity of an anti-p24 (HIV-1) monoclonal antibody. Cell 91:799–809

Liu Z, Song D, Kramer A, Martin A, Dandekar T, Schneider-Mergener J, Bautz EKF, Dübel S (1999) Fine mapping of the antigen-antibody interaction of scFv217, are combinant antibody inhibiting RNA polymerase from *Drosophila melanogaster*. J Mol Recognit 12:103–111

Mahler M, Mierau R, Blüthner M (2000) Fine specificity of the B-cell anti-CENP-A autoimmunresponse. J Mol Med 78:460–467

Moroi Y, Peebles C, Fritzler MJ, Steigerwald J, Tan EM (1980) Autoantibody to centromere (kinetochore) in scleroderma sera. Proc Natl Acad Sci USA 77:1627–1631

Muro Y, Iwai T, Ohashi M (1996) A charged segment mainly composed of basic amino acids forms an autoepitope of CENP-A. Clin Immunol Immunopathol 78:86–89

Muro Y, Azuma N, Onouchi H, Kunimatsu M, Tomita Y, Sasaki M, Sugimoto K (2000) Autoepitopes on autoantigen centromere protein-A (CENP-A) are restricted to the N-terminal region, which has no homology with histone H3. Clin Exp Immunol 120:218–223

Reineke U, Sabat R, Welfle H, Volk H-D, Schneider-Mergener J (1999) A synthetic mimic of a discontinuous binding site on interleukin-10. Nature Biotechnol 17:271–275

Sullivan KF, Hechenberger M, Masri K (1994) Human CENP-A contains a histone H3 related histone fold domain that is required for targeting to the centromere. J Cell Biol 127:581–592

Tan EM, Rodnan GP, Garcia I, Moroi Y, Fritzler MJ, Peebles C (1980) Diversity of antinuclear antibodies in progressive systemic sclerosis. Anti-centromere antibody and its relationship to CREST syndrome. Arthritis Rheum 23:617–625

Suppliers

Sigma-Genosys Ltd.
(e-mail: info@sigma-genosys.co.uk,
Internet: http://www.Genosys.co.uk/,
Tel.: +44-1223-839000, Fax: +44-1223-839300)
UK/Europe, London Road, Pampisford, Cambridge CB2 4EF, UK

Amersham Pharmacia Biotech
(e-mail: cust.orderde@eu.apbiotech.com,
Internet: http://www.apbiotech.com/na/,
Tel.: +49-800-9080713, Fax: +49-800-9080714)
Munzingerstrasse 9, 79111 Freiburg, Germany

Protein-Protein Interactions

MATTHEW R. GROVES and IRMGARD SINNING

Introduction

The characterization of the manner in which specific proteins interact is of immense interest not only for basic research but also for therapeutic research purposes. The identification of bioactive peptides can lead to new drug and therapeutic leads in biological systems in which proteins interact (e.g. signal transduction, protein targeting, cell cycle control). The identification of interacting peptides within multi-protein assemblies can also aid in the design of further experiments (e.g. mutagenesis, functional competition with peptides/domains) or the generation of computer models of the manner in which the proteins interact.

Folded proteins interact through surface-exposed hydrophobic and hydrophilic amino acids. These solvent-exposed residues may contribute to a single or a number of distinct protein-protein interaction surfaces, and the vast majority of proteins fold such that amino acids which contribute to the same surface are distantly separated in the primary sequence. Thus, the identification of a single interacting region in the primary sequence of a protein may often not be the complete picture, and techniques that allow the entire primary sequence of a protein to be simultaneously probed for interacting peptides enable the rapid and efficient identification of protein-protein interaction sites.

M.R. Groves, I. Sinning (✉) (e-mail: Irmi.Sinning@bzh.uni-heidelberg.de, Tel: +49-6221-544780, Fax: +49-6221-544790)
Structures Programme, EMBL, Meyerhofstrasse 1, 69117 Heidelberg, Germany
Current address: Dept. of Structural Biology, BZH, University of Heidelberg, INF 328, 69120 Heidelberg, Germany

Springer Lab Manual
J. Koch, M. Mahler (Eds.) Peptide Arrays on Membrane Supports

Only recently have immobilized peptide scans been used as a powerful tool to identify peptides important in the formation of protein-protein complexes (Frank 1992; McCarty et al. 1996; Hilpert et al. 1999; Knoblauch et al. 1999; Piossek et al. 1999; Schultz et al. 1999; Brix et al. 2000; Groves et al. 2001). The technique is essentially an extension of the one used to map antibody epitopes, which is described in Chapter 5 of this manual (see also Korth et al. 1997; Kramer et al. 1997). Entire protein sequences may be synthesized on cellulose membranes and probed with a known or potential interacting protein or proteins. Through the use of a pipetting robot, entire multi-protein assemblies may be examined for interaction with a probe protein or proteins in a coarse fashion (3- or 5-amino-acid offset between synthesized peptides). Alternatively, once an interacting peptide is identified, probing an alanine scan (in which each amino acid of the peptide is replaced by an alanine residue; see Chap. 9) of a bioactive peptide can provide highly detailed data on the contribution of each amino acid to the interaction.

However, as the interactions between a protein and the individual peptides of its binding partner(s) are, in general, much weaker than those between antibodies and their epitopes, certain additional precautions must be taken:

- Firstly, the experiment must be repeated in number of incubation buffers so that non-specific interactions may be identified.
- Secondly, as a consequence of the need to repeat the experiment, care must be taken to ensure that the immobilized peptide scan membranes are entirely free of protein after regeneration of the membrane and before incubation with fresh probe protein.

Bound material may be detected in a number of ways. Described below in the standard procedure is detection through the use of HRP-conjugated antibodies in combination with a chemiluminescent reagent. However, the use of streptavidin-HRP complexes in combination with biotinylated probe proteins is just as sensitive. Further, the use of radiolabeled proteins derived from an in vitro translation system and/or the transfer of any bound proteins to a nitrocellulose membrane can increase the sensitivity of the procedure by reducing the number of washing steps. Although the technique is sensitive, any data obtained should be confirmed either through in vitro (or in vivo) mutation/deletion analysis or other biochemical assays. Ideally, structural analysis of protein-protein complexes or protein-peptide complexes would provide confirmation of a predicted interaction surface between two proteins.

Outline

The peptides corresponding to the sequence of a protein are synthesized on cellulose membrane as described in Chapters 3 and 4. These membranes are first incubated with the antibody detection system in order to determine the background signal obtained from the detection system alone. If the background is acceptably small, the membrane is then incubated with the probe protein. Detection is either carried out through direct immunodecoration of bound probe protein on the peptide membrane or after transfer to a nitrocellulose membrane. The membranes are then regenerated before the experiment is repeated under different incubation conditions. For a general outline see Fig. 1.

Materials

Materials required for protein incubation:

- Immobilized oligopeptides of the protein(s) of interest (see Chaps. 3 and 4).
- Purified probe protein(s) that interact with the protein spotted in form of peptides on the membrane.

  Materials required for the detection of bound proteins:
- Specific primary antibody against the probe protein and HRP-conjugated secondary antibody.
- Chemoluminescent detection system (e.g. ECL Western blotting detection reagents; Amersham Pharmacia Biotech, Piscataway, NJ, USA)
- Photographic film (e.g. Hyperfilm Amersham Pharmacia Biotech)
- Developing apparatus (e.g. Amersham Lifescience Hyperprocessor, Amersham Pharmacia Biotech, Piscataway, NJ, USA)

General Outline of Experimental Procedure

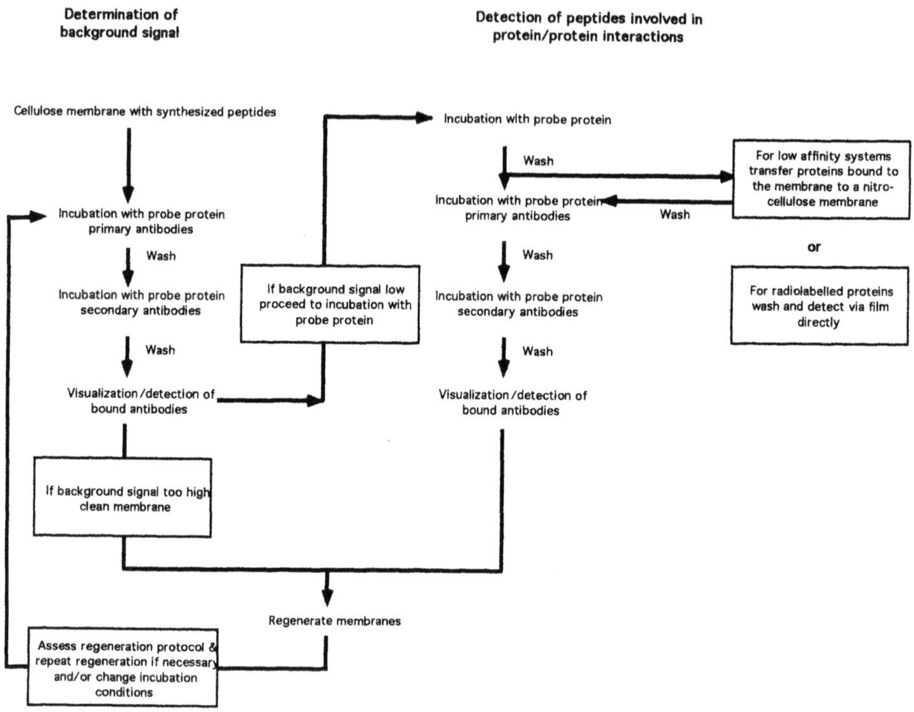

Additional materials for lower affinity protein-protein interactions:

- Nitrocellulose transfer-membrane (e.g. Protan, Schleicher and Schuell, Germany)
- Blotting paper (e.g. Machery-Nagel, Dueren, Germany, MN 440 B)
- Blotting apparatus (e.g. Biometra-Fast-Blot, Göttingen, Germany, B337593)

Buffers The incubation buffer should be based on conditions under which the protein of interest is stable in its native conformation. Different buffers, salt concentrations, pH or additional components (e.g. cofactors, lipids or detergents) may be required.

Fig. 1. The general experimental outline. The membranes are initially probed with the detection system in the absence of probe protein in order to discern the level of background obtained from the detection method. The membranes are then incubated with the probe proteins before detection. Under the standard procedure, the membranes are sequentially incubated with the primary and secondary antibodies before horseradish peroxidase (HRP)-conjugate chemiluminescent detection. For low-affinity interactions, probe protein bound to the membranes may either be transferred to a nitrocellulose membrane and immobilized before detection or detected directly through the use of radiolabeled in-vitro-translated material

- Incubation buffer (TBS, Tris-buffered saline):
  10 mM Tris/HCl (pH 7.6)
  150 mM NaCl
- Optional additional blocking components:
  0.02 % (v/v) Tween-20 and/or
  0.5 % (w/v) Milk powder
- Blocking buffer:
  TBS
  0.5 % (w/v) Skimmed milk powder
  (For other blocking conditions see Chap. 11).
- Strip buffer A:
  50 % Ethanol
  10 % Glacial acetic acid
- Strip buffer B:
  50 mM Tris/HCl (pH 8.0)
  10 mM β-mercaptoethanol
  10 % (w/v) Sodium dodecyl sulfate (SDS)
- Transfer Buffer for western blotting:
  25 mM Tris/HCl (pH 7.6)
  192 mM glycine
  20 % Methanol
  0.03 % SDS

Procedure

Standard procedure

In general, the following procedure is sufficient to pick up protein-protein interactions that stabilize large multi-protein complexes and should be the general starting point of an investigation using membrane-immobilized peptides. However, if this should yield little (or no) data, or the interactions are known to be weak, then a second procedure optimized for low-affinity interactions is described below.

Determination of background signal due to non-specific binding

1. Incubate the membrane with the primary antibody in 10 ml of the incubation buffer with gentle agitation for 1 h. **Note:** The incubation buffer should be kept constant throughout the rest of the experiment, so it is advisable to prepare a large volume (200 ml). The concentration of primary and secondary antibodies required for a good final signal will vary from antibody to antibody, but should be kept constant within each experiment. Typically, secondary antibodies are diluted as recommended for immunoblotting (see manufacturer's instructions) ranging from 1000- to 5000-fold dilutions (around 1 μg/ml). Depending on the primary and secondary antibodies, an initial blocking step may be required in order to decrease the non-specific binding of antibodies to the cellulose membrane (incubation buffer + 0.5–2% (w/v) skimmed milk powder for 2 h at room temperature or overnight at 4 °C on a side-to-side shaker).

2. Discard the incubation buffer containing the primary antibody and wash the membrane three times in fresh incubation buffer for 10 min. The wash should be discarded each time.

3. Incubate the membrane with the secondary HRP-conjugated antibody in 10 ml incubation buffer with gentle agitation for 1 h.

4. Discard the incubation buffer containing the secondary antibody and wash the membrane three times in fresh incubation buffer for 10 min.

5. Detect any bound secondary antibody using a chemiluminescent reagent and an imaging film following the manufacturer's instructions. **Note:** A number of exposure times should be chosen (e.g. 1, 5, 10 and 30 min). If no signal is detected on the membranes, there is assumed to be no background contribution from the primary or secondary antibodies of the detection system under the current

incubation conditions and steps 6 and 7 below should be omitted.

Regeneration of the membrane

6. If any signal is detected, remove bound antibodies from the membrane by incubation with 10 ml of the strip buffer(s). **Note:** Generally, alternating 30-min incubations in strip buffers A and B three times are sufficient to remove any bound protein. However, some bound protein may remain and harsher conditions may need to be investigated in order to clean the membranes (e.g. increase in temperature, high pH, 500 mM lithium perchlorate).

7. Repeat steps 3–6 above in order to evaluate the efficiency of the stripping protocol. If an acceptably low signal is obtained – indicating that the membranes are free of primary antibody – repeat steps 1–6 under different incubation conditions (e.g. higher salt, lower amounts of primary and/or secondary antibodies) until an acceptably low background signal for both primary and secondary antibodies is obtained, then proceed to step 8 below. If a high signal is still retained at this point, then harsher stripping conditions and/or an alternative, more specific secondary antibody may be required.

Protein-peptide interaction

8. Incubate the stripped membrane (from step 5 or 7) with pure probe protein at concentrations of 5–200 nM in the incubation buffer. **Note:** It is advisable to start with lower concentrations of protein and repeat the experiment(s) with higher concentrations, as lower amounts of probe protein are easier to remove and higher amounts of protein may give rise to non-specific signals.
9. Discard the incubation buffer containing the probe protein and wash the membrane three times in fresh incubation buffer for 10 min each time.
10. Perform detection of bound probe protein according to steps 1–5 above.
11. Regenerate the membrane as described above (see step 6). However, in order to detect any residual probe protein it is necessary to immunodecorate again with the primary and secondary antibodies following steps 1–6 above. Repeat steps 1–6 until the signal obtained from the stripped membranes is

at background levels. **Note:** Some additional modification of the stripping protocol obtained in step 6 may be required in order to regenerate the membrane completely.

12. The protocol may then be repeated with variations in probe protein-antibody concentration and incubation buffer in order to optimize the signal obtained.

Lower-affinity protein–protein interaction

Low-affinity interactions may not be sufficiently strong to survive the numerous washing steps described in the above procedure. The following protocol allows detection of interacting peptides using fewer wash steps before any interacting probe protein is immobilized and/or detected. This procedure is only useful in cases in which the standard procedure has failed to yield data. In cases in which the standard procedure has given a signal, there is a danger of picking up non-specific interactions with this method.

1. Incubate the membranes with the probe protein analogous to step 8 above.
2. Discard the incubation buffer containing the probe protein and wash the membrane once in fresh incubation buffer for 10 min.
3. Briefly equilibrate (3–5 min) both the peptide membrane and a sheet of nitrocellulose trimmed to fit the peptide membrane in transfer buffer for Western blotting.

Electrotransfer of bound probe protein

4. Electrotransfer the proteins bound to the immobilized peptides onto nitrocellulose for 1 h using 0.85 mA per square centimeter of membrane surface. Due to the denaturation by SDS, all proteins should have acquired a negative charge. Therefore, the nitrocellulose should be placed on the positive-electrode side of the immobilized peptide membrane. **Note:** Depending on the chemical properties of the probe protein and its affinity to the immobilized peptides, the time required for the transfer might differ and should be determined empirically. Ensure that the antibody detection system recognizes the denatured probe protein.
5. Block the nitrocellulose membrane with 2% (w/v) milk powder in TBS for 2 h at room temperature.

6. Proceed with the detection of electrotransferred proteins following steps 1–5 of the standard procedure.
7. The immobilized peptide membrane may then be regenerated following step 6 of the standard procedure and the regeneration protocol assessed following steps 4–6 above.

Results

Although protein-protein interaction surfaces are frequently contributed to by a number of discontinuous peptides, the individual contributions from peptides may nonetheless be detected. cpSRP is a signal recognition particle (SRP) -like targeting system responsible for the transport of the highly abundant light-harvesting membrane protein LHC II from the inner chloroplast membrane to the thylakoid membranes. Studies have shown that a targeting complex consisting of cpSRP43 and cpRP54 solubilize and traffic the LHC II membrane protein in a GTP-dependent manner (Franklin and Hoffman 1993; Hoffman and Franklin 1994; Li et al. 1995; Schuenemann et al. 1998; Klimyuk et al. 1999). In order to examine the amino acids essential in forming this heterotrimeric protein-protein complex, peptides corresponding to the complete LHC II sequence were synthesized on cellulose membrane and probed with highly purified cpSRP components both alone and in complex with each other (Groves et al. 2001). The immobilized peptide scan data shown in Fig. 2 clearly show the individual preference of the components of the cpSRP targeting complex (Fig. 2b). cpSRP54 is clearly detected in spots which contain contributions from the three hydrophobic transmembrane helices of LHC II, consistent with the high (>74%) sequence homology it shares with other members of the SRP54 family and their known preference for hydrophobic signal sequences (Krieg et al. 1986; reviewed in Lütcke 1995). However, larger amount of cpSRP54 are detected in spots that contain contributions from the third transmembrane helix (TM3), in agreement with the findings of High et al. (1997) that TM3 is the targeting signal of LHC II. cpSRP43 is also readily detected in peptides that contain amino acids from a hydrophilic region N-terminal to the third transmembrane helix L18, again consistent with previously published data (DeLille et al. 2000) and also in spots that contain residues from the TM3. The

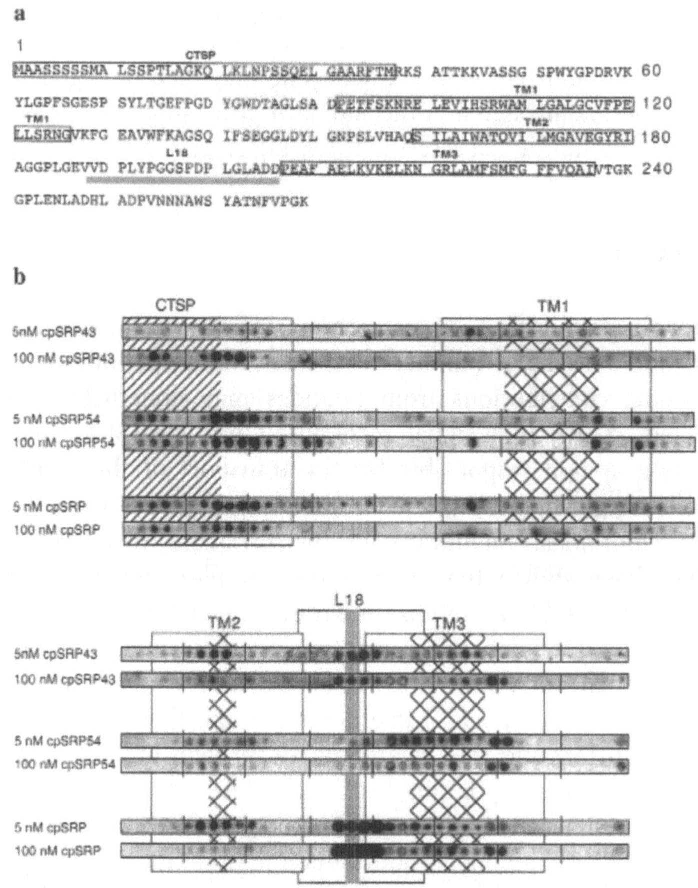

Fig. 2. **a** The sequence of chloroplast pre-LHC II used in the immobilized peptide scan experiment. The three hydrophobic transmembrane helices (TM1, TM2 and TM3) and the chloroplast signal transit peptides (CTSP) are highlighted in *enclosed boxes*, the hydrophilic L18 sequence N-terminal to TM3 is *underlined*. **b** Immobilized peptide scans (18-mers with a 3-amino-acid shift) of the chloroplast membrane protein pre-LHC II were incubated with components of the cpSRP (signal recognition particle) targeting system both alone (rows labeled *cpSRP43* and *cpSRP54*) and in complex (row labeled *cpSRP*). Membranes probed with 5 nM of probe protein were incubated in incubation buffer (see text) with an additional 0.02% (v/v) Tween-20. Membranes incubated with 100 nM probe protein were incubated in incubation buffer with an additional 0.5% (w/v) milk powder. Detection was carried out as described in the text following the standard procedure. Peptide spots that are comprised entirely of amino acids within the CTSP, L18 or transmembrane helices are highlighted by *filled diagonally hatched*, *gray* or *cross-hatched* boxes, respectively. Peptides that contain contributions from the CTSP, L18 or transmembrane helices are indicated in the *boxes* drawn around the respective *filled boxes*. Note the overlap between the L18 and TM3 boxes. Figure taken from Groves et al. (2001) copyright JBC, available from JBC online (http://www.jbc.org)

increased signal obtained when the same membranes are probed with the cpSRP54/43 complex indicates that both proteins are involved in the formation of the heterotrimer protein complex: cpSRP54 through an interaction with the TM3 of LHC II and cpSRP43 through an interaction with the L18 and TM3 of LHC II.

The spots C-terminal to the L18 motif, which also contain contributions from TM3, show large amounts of bound cpSRP complex, whereas spots N-terminal to the L18 motif do not. This indicates that L18 and TM3 are two parts of a discontinuous interacting site with both proteins. The increased signal obtained also indicates that both proteins contribute to the binding of the L18TM3 motif. The interaction of the cpSRP complex components with the chloroplast signal transit peptide (CTSP) may also have implications for the import of LHC II into chloroplasts.

The differences in the signal obtained by the variation of the incubation conditions suggest that the signal for the first two transmembrane helices is non-specific.

Troubleshooting

- The incubation buffer should be one in which the probe protein is stable. Additional components may be necessary (e.g. detergent, co-factors).
- The membrane should be **completely** immersed in liquid during all steps.
- The background signal from the primary and secondary antibodies should be determined first to avoid confusion between background signal and that from any material not removed by the stripping protocol.
- Start with lower concentrations of protein to incubate with the membrane, as the membranes are easier to regenerate.
- The antibodies/probe protein may be incubated using small volumes in order to conserve material. Placing the membrane in a clean box and covering with Parafilm allows incubation with volumes of 1–2 ml.
- As the chemiluminescent reagent degrades spontaneously, reapplication after long exposures will improve the measured signal.

- An over-exposed membrane is useful in order to identify the positions of each of the spots in the scan.
- Poly-acidic (DDDDDDDD, poly-basic (RRRRRRRR) and poly-hydrophobic (FFFFFFFFF) spots on the membrane allow the evaluation of the background signal expected from non-specific interaction of the probe.

References

Brix J, Ziegler GA, Dietmeier K, Schneider-Mergener J, Schulz GE, Pfanner N (2000) The mitochondrial import receptor tom70: identification of a 25 kDa core domain with a specific binding site for preproteins. J Mol Biol 303:479–88

DeLille J, Peterson EC, Johnson T, Moore M, Kight A, Henry R (2000) A novel precursor recognition element facilitates posttranslational binding to the signal recognition particle in chloroplasts. Proc Natl Acad Sci USA 97:1926–31

Frank R (1992) SPOT synthesis: an easy technique for the positionally addressable, parallel chemical synthesis on a membrane support. Tetrahedron 48:9217–9232

Franklin AE, Hoffman NE (1993) Characterisation of a chloroplast homologue of the 54-kDa subunit of the signal recognition particle. J Biol Chem 268:22175–22180

Groves MR, Mant A, Kuhn A, Koch J, Dubel S, Robinson C, Sinning I (2001) Functional characterization of recombinant chloroplast signal recognition particle. J Biol Chem 276:27778–86

High S, Henry R, Mould RM, Valent Q, Meacock S, Cline K, Gray JC, Luirink J (1997) Chloroplast SRP54 interacts with a specific subset of thylakoid precursor proteins. J Biol Chem 272:11622–11628

Hilpert K, Behlke J, Scholz C, Misselwitz R, Schneider-Mergener J, Höhne W (1999) Interaction of the capsid protein p24 (HIV-1) with sequenced-derived peptides: influence on p24 dimerization. Virology 254:6–10

Hoffman NE, Franklin AE (1994) Evidence for a stromal GTP requirement for the integration of a chlorophyll a/b-binding polypeptide into thylakoid membranes. Plant Physiol 105:295–304

Klimyuk VI, Persello-Cartieaux F, Havaux M, Contard-David P, Schuenemann D, Meiherhoff K, Gouet P, Jones JD, Hoffman NE, Nussaume L (1999) A chromodomain protein encoded by the *Arabidopsis* CAO gene is a plant-specific component of the chloroplast signal recognition particle pathway that is involved in LHCP targeting. Plant Cell 11:87–99

Knoblauch NT, Rudiger S, Schonfeld HJ, Driessen AJ, Schneider-Mergener J, Bukau B (1999) Substrate specificity of the SecB chaperone. J Biol Chem 274:34219–25

Korth C, Stierli B, Streit P, Moser M, Schaller O, Fischer R, Schulz-Schaeffer W, Kretzschmar H, Raeber A, Braun Ehrensperger F, Hornemann S, Glocks-

huber R, Riek R, Billeter M, Wuthrich K, Oesch B (1997) Prion (PrPSc)-specific epitope defined by a monoclonal antibody. Nature 390:74–77
Kramer A, Keitel T, Höhne W, Schneider-Mergener J (1997) Molecular basis of binding promiscuity of an anti-p24 (HIV-1) monoclonal antibody. Cell 91:799–809
Krieg UC, Walter P, Johnson AE (1986) Photocrosslinking of the signal sequence of nascent preprolactin to the 54-kilodalton polypeptide of the signal recognition particle. Proc Natl Acad Sci USA 83:8604–8
Li X, Henry R, Yuan J, Cline K, Hoffman NE (1995) A chloroplast homologue of the signal recognition particle subunit SRP54 is involved in the posttranslational integration of a protein into thylakoid membranes. Proc Natl Acad Sci USA 92:3789–93
Lütcke H (1995) Signal recognition particle (SRP), a ubiquitous initiator of protein translocation. Eur J Biochem 228:531–50
McCarty J, Rüdiger S, Schönfeld HJ, Schneider-Mergener J, Nakahigashi K, Yura T, Buckau B (1996) Regulatory region C of the *E. coli* heat shock transcription factor, sigma32, constitutes a DnaK binding site and is conserved among eubacteria. J Mol Biol 256:829–837
Piossek C, Schneider-Mergener J, Schirner M, Vakalopolou E, Germeroth L, Thierauch KH (1999) Vascular endothelial growth factor (VEGF) receptor II-derived peptides inhibit VEGF. J Biol Chem 274:5612–5619
Schuenemann D, Gupta S, Persello-Cartieaux F, Klimyuk VI, Jones JDG, Nussaume L, Hoffman NE (1998) A novel signal recognition particle targets light-harvesting proteins to the thylakoid membranes. Proc Natl Acad Sci USA 95:10312–10316
Schultz J, Hoffmüller U, Ashurst J, Krause G, Schmieder PJ, Macias M, Schneider-Mergener J, Oschkinat H (1998) Specific interactions between the synthrophin PDZ domain and voltage-gated sodium channels. Nature Struct Biol 5:19–24

Suppliers

Amersham Pharmacia Biotech
(e-mail: cust.orderde@eu.apbiotech.com,
http://www.apbiotech.com/na/,
Tel.: +49-800-9080713, Fax: +49-800-9080714)
Munzingerstrasse 9, 79111 Freiburg, Germany

Sigma-Aldrich Chemie GmbH
(e-mail: deorders@eurnotes.sial.com,
http://www.Sigma-aldrich.com/,
Tel.: +49-89-65130, Fax: +49-89-65131169)
Eschenstrasse 5, 82024 Taufkirchen, Germany

Macherey-Nagel GmbH and Co. KG
(e-mail: sales@macherey-nagel.de,
http://www.macherey-nagel.com/,
Tel.: +49-2421-9690, Fax: +49-02421-969199)
Postfach 10 13 52, 52313 Düren, Germany

Analysis of Protein-DNA Interactions

MONIKA REUTER and ELISABETH MÖNCKE-BUCHNER

Introduction

By analyzing specific protein-DNA interactions, molecular recognition processes can be better understood. How do distinct proteins recognize specific DNA sequences? Prokaryotic restriction endonucleases, for example, are characterized by highly specific DNA interactions and therefore are excellent models to study recognition events. More than 3,300 type II restriction endonucleases are known, recognizing about 200 distinct DNA specificities (Roberts and Macelis 2001). Among this large group of functionally related enzymes is a considerable number of isoschizomers, proteins isolated from different microbial species that interact with the same DNA sequence. However, there does not appear to be noteworthy homology in amino acid sequence among these enzymes (for a review see Pingoud and Jeltsch 2001). Did enzymes that interact with the same DNA sequence develop independent structures or mechanisms to interact with a specific target site? Are there any as yet unknown common rules of DNA recognition? To date, X-ray crystallography of protein-DNA complexes is without doubt the most informative method for understanding specific recognition. However, X-ray crystallography can be slow because obtaining good diffracting crystals is a very time-consuming and unpredictable process. Thus, faster methods are sought to increase the data pool on recognition complexes.

M. Reuter (✉) (e-mail: monika.reuter@charite.de,
Tel.: +49-30-450-525201, Fax: +49-30-450-525907),
E. Möncke-Buchner
Institut für Virologie, Medizinische Fakultät der Humboldt-Universität (Charité), 10098 Berlin, Germany

Springer Lab Manual
J. Koch, M. Mahler (Eds.) Peptide Arrays on Membrane Supports

Numerous specific protein-DNA interactions were confirmed when they were reconstructed at the peptide-oligonucleotide duplex level. In these experiments soluble synthetic peptides of different length were used to show that a particular short amino acid sequence is involved in a specific DNA interaction. This could be shown directly when the synthetic peptides bound DNA or indirectly when they competed with the native protein at the binding site. Examples where soluble peptides have been successfully applied are the yeast transcriptional activator GCN4 (Talanian et al. 1990, 1992), the human DNA-binding protein hRFX1 (Cornille et al. 1998), the prokaryotic restriction endonuclease *Eco*RI (Jeltsch et al. 1995), and the RecA protein (Wang et al. 1998).

Recently, we have adapted the peptide-scan method to determine protein-DNA contacts. With this method we have effectively used peptides that are membrane-bound at their C-terminal end instead of soluble peptides to search for regions involved in specific DNA binding (Reuter et al. 1999). These peptides covalently bound to the cellulose membrane are overlapping and span the complete amino acid sequence of a protein. The peptide-scan method offers the possibility to screen a large number of peptides at once regarding their ability to bind DNA. Thus, for optimal results on identifying DNA-binding sites in a protein, we recommend applying the peptide-scan method as the main component of a wider mapping strategy together with site-specific mutagenesis, DNA-protein cross-linking experiments, and/or limited proteolysis.

Outline

A flow diagram of a typical DNA-peptide binding experiment is shown in Fig. 1.

Materials

Equipment

- Phosphorimager (e.g. PhosphorImager SI, Amersham Pharmacia Biotech)
- ImageQuant software (Amersham Pharmacia Biotech)

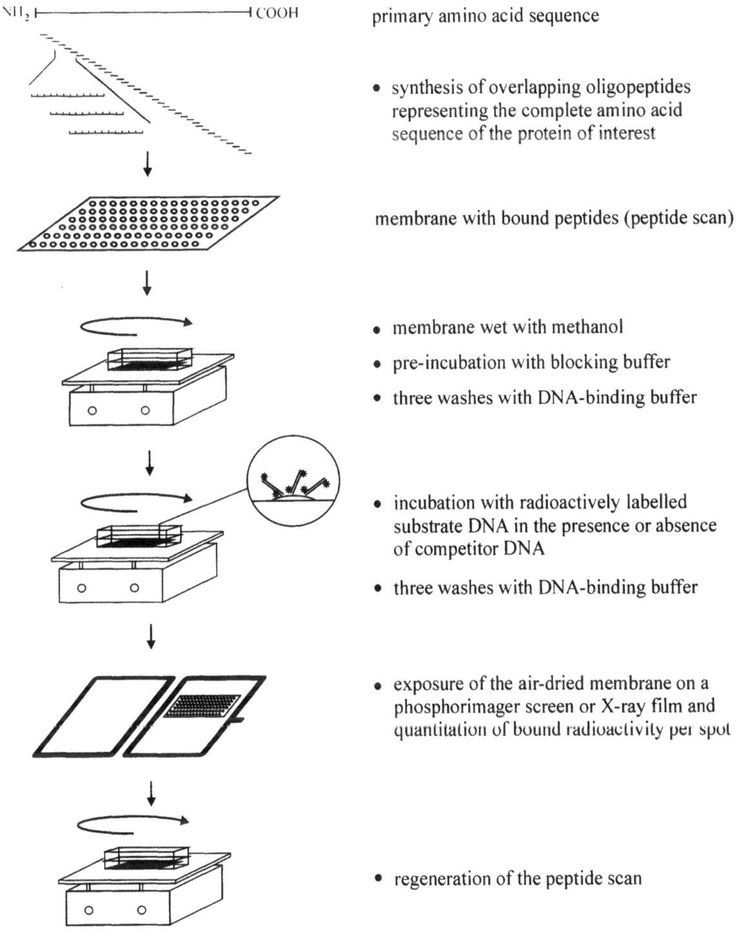

Fig. 1. Mapping DNA-binding regions of a protein using membrane-bound synthetic oligopeptides – a flow chart

- Blocking buffer:
  100 mM maleic acid
  150 mM NaCl, pH 7.5
  1 % Blocking reagent (Roche Biochemicals)
- DNA-binding buffer:
  33 mM TrisOAc
  66 mM KOAc
  10 mM $Mg(OAc)_2$, pH 7.6
- Strip buffer A:
  DNA-binding buffer
  2 M NaCl

- Strip buffer B:
  1 M K_2HPO_4, pH 8.0

Procedure

Buffer composition In general, the DNA-binding buffer should be based on the conditions under which the protein is active. DNA-binding buffer should be prepared with and without known cofactors of the protein of interest to test for their relevance in the specific protein-DNA interaction. In our model system, 10 mM $Mg(OAc)_2$ was added to the binding buffer, as Mg^{2+} is an essential cofactor for the catalytic activity of native restriction endonucleases. The presence of magnesium ions led to a pronounced increase of DNA-binding specificity.

Preparation of the peptide scan The following steps are performed in polyacryl or polypropylene vessels that are shaken gently at room temperature:

1. Place the membrane in an appropriate vessel using plastic forceps.
2. Wash the membrane in a small volume of methanol by shaking it gently for 5 min at room temperature.
3. Remove the methanol carefully.
4. Pre-incubate the membrane in blocking buffer for 1 h to avoid unspecific binding to the membrane.
5. Wash the membrane three times for 10 min each with DNA-binding buffer.

DNA-peptide interaction

6. Incubate the peptide scan with a defined amount (e.g. 25–50 pmol) of a ^{32}PγATP-labeled specific DNA probe in a total volume of 10 ml DNA-binding buffer for 30–60 min at room temperature. Usually, oligonucleotide duplexes 16–30 bp long are sufficient to interact with the peptides. **Note:** Alternatively, non-radioactive detection systems such as biotinylated DNA probes in combination with streptavidine-HRP-complexes may be used.

Control experiments To prove whether the interaction of the radioactively labeled DNA substrate is specific, competition experiments should be

performed. Usually we run three experiments, either in parallel with three identical membranes or successively with one (regenerated) peptide scan. Either of the following is added to the membranes: (a) radioactively labeled substrate DNA alone; (b) same as (a) but with saturating amounts of unlabeled specific DNA or (c) same as (a) but with saturating amounts of unlabeled unspecific DNA. Since DNA length and sequence may influence DNA binding, the unspecific competitor should have the same length and overall base composition as the specific one. In addition, if possible, one should omit the specific DNA recognition sequence on the dinucleotide level.
For a saturating competitor, the competing molecules should be in molar excess over the sum of peptides on the membrane. Otherwise, it is conceivable that the binding capacity of a peptide spot is high enough to trap both specific and unspecific DNA and thus no competing effects are detected.

7. Wash the membrane three times for 5 min each with 10 ml of DNA-binding buffer. The length of time of the washing steps should be fixed and depends on the stability of DNA-peptide binding obtained.

8. Allow the membrane to air-dry.

Visualization and quantitation of the bound DNA

9. Expose the membrane on an X-ray film or to the phosphorimager screen.

An experimental control has to be included to compare bound radioactivity quantitatively for a distinct spot in different experiments. We used a serial dilution of ^{14}C isotope in each experiment as a standard. Absolute radioactivity values per spot were referred to 1,000 counts of the ^{14}C standard. When comparing the effect of competitor DNA in consecutive experiments, the absolute amount of radioactivity used in each experiment has to be taken into account.

Regeneration of the membrane

10. Peptide scans can be stripped by extensive washing (several days with regular buffer changes) in Strip buffer A or B.

11. Dried membranes were stored at –20 °C.

Results

The following figures show the results of a typical DNA-binding experiment after visualization and quantitation with a phosphorimager [adopted from Reuter et al. (1999)]. The peptide spots that interact with the radioactively labeled substrate DNA in the absence of Mg^{2+} ions are focused to a few distinct regions in the primary amino acid sequence (Fig. 2a). These peptides show a high affinity to the oligonucleotide duplex. However, the presence of Mg^{2+} ions clearly diminishes unspecific DNA binding and points out two very specific DNA-binding regions (Fig. 2b). The quantitative influence of the cofactor is evident in Fig. 2 c if one compares the data from Fig. 2a with those of Fig. 2b.

To determine whether the DNA-peptide binding was specific for the DNA target sequence, binding of oligonucleotide duplexes

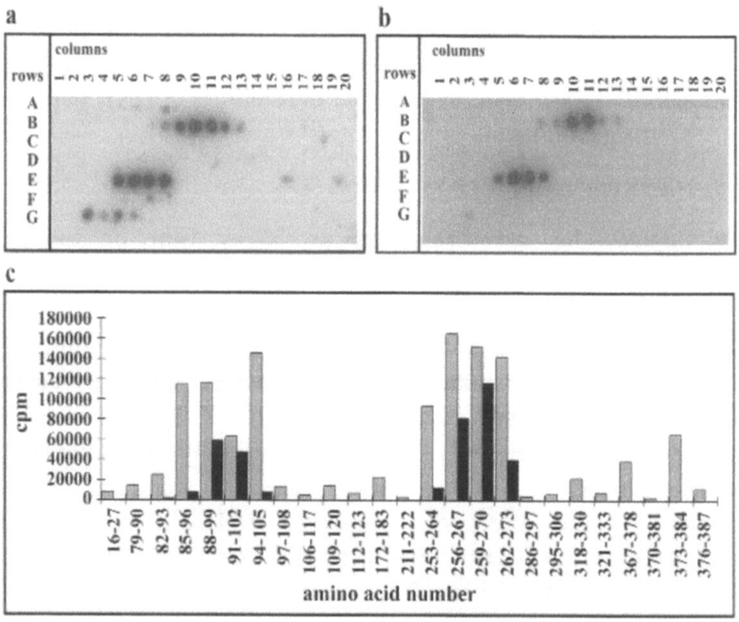

Fig. 2a–c. DNA binding to a synthetic *Eco*RII peptide scan comprising the complete amino acid sequence. A cellulose membrane with 132 *Eco*RII-derived dodecapeptides (neighboring peptides overlap by 9 amino acids) was incubated with a radioactively labeled specific oligonucleotide duplex in the absence (**a**) and in the presence (**b**) of 10 mM Mg^{2+}. **c** The quantitative differences in the presence (*black*) or absence (*gray*) of magnesium are shown. The N-terminal peptide of the protein is at position A1, and the C-terminal peptide is at G12

under competitive conditions was done (Fig. 3). Usually, the amount of the competing oligonucleotide duplexes was sufficient to saturate the total peptide amount in the spots. The unlabeled substrate was in 10^4- to 10^5-fold molar excess of the labeled substrate. Data of the competition experiments were analyzed quantitatively using a ^{14}C standard. Under strict competitive conditions, five peptide spots specifically bound oligonucleotides containing the target sequence (Fig. 3). These spots comprise *Eco*RII amino acid positions 88–102 and 256–273. In contrast to unlabeled unspecific oligonucleotides, unlabeled specific oligonucleotide duplexes chased the binding of the labeled specific oligonucleotides to below 1 %. Since the labeled unspecific competing oligonucleotide duplex itself bound to these *Eco*RII peptides with an efficiency between 1 and 10 %, we strongly recommend competition experiments as a specificity assay.

After having found peptide spots involved specifically in DNA binding, substitution analogues of the potential binding regions were tested in which every amino acid of the original sequence was replaced by all 20 natural occurring amino acids (see also Chap. 9). Incubation of the membrane containing the substitution analogues followed the procedure described above.

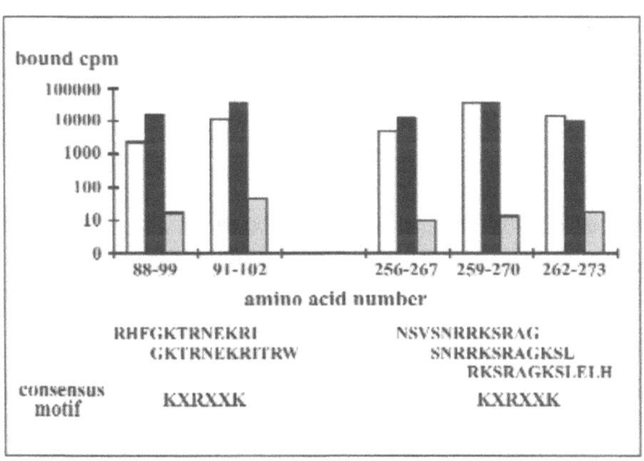

Fig. 3. Binding of a specific substrate under competitive conditions. DNA binding was performed in the absence (*white*) and in the presence of saturating amounts of unspecific (*black*) or sequence-specific (*gray*) competitor DNA. The peptide sequences of the spots that bound the substrate, as well as the minimal consensus motif present in both amino acid sequences, are indicated at the *bottom*

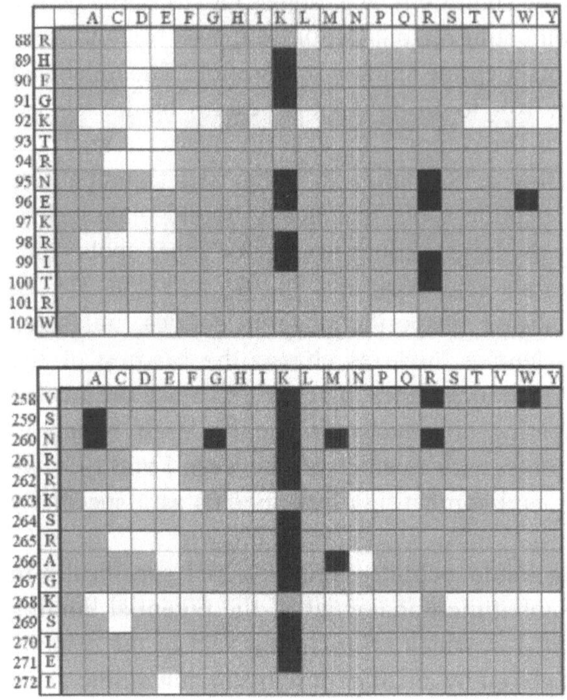

Fig. 4. Influence of single amino acid substitution on binding of *Eco*RII-specific DNA sequences. The amino acid sequences and positions of both potential binding regions are indicated at the *left*. Substituents at the respective position are indicated horizontally. Binding strength within the three-fold standard deviation range calculated from DNA binding to all spots of unsubstituted *Eco*RII peptides is shown in *gray*. Substitutions that decreased DNA-binding are represented as *white squares*; increased DNA binding is shown in *black*. *Upper panel* binding site I, *lower panel* binding site II

Figure 4 (Reuter et al. 1999) displays the layout of a substitution scan. The original amino acid sequence of the identified DNA-binding regions occurs twice in each row: in the column on the left and in the column of the substituent that reconstitutes the original sequence. To evaluate how single amino acid substitutions affect the DNA-binding efficiencies of a given peptide, we calculated the initial binding efficiency as the average of all original peptide spots on the membrane plus or minus the three-fold standard deviation. Binding values outside this range were considered significant.

Comments

One has to take into account the following points for the peptide scan method:

- If the complete amino acid sequence of a protein is split into single peptides that are fixed with one terminus to a membrane, and thus these peptides can interact with oligonucleotide duplexes, binding affinity may be several orders of magnitude less than that of native three-dimensional structures.
- It is conceivable that peptide scans emphasize the contribution of electrostatic binding at the expense of other interactions. For this reason, the specificity of detected DNA-peptide interaction has to be proven.
- Besides the competition experiments described above, the significance of identified DNA-binding peptides can be further verified by quantifying DNA binding after scrambling of their amino acid residues. This can also be done with membrane-bound peptides of the original and scrambled sequence. We would like to emphasize that, for quantitative evaluation, the respective peptides should be present in multiple copies on the membrane.
- Experimental data are most reproducible when a moderate peptide density per spot (approximately 0.6 nmol peptides per spot) is used and when a minimal distance between the spots is kept. Otherwise, unspecific interactions may occur that may be due to unfavorable charge effects between the high peptide amount and the substrate DNA.

Acknowledgements. We would like to thank P. Mackeldanz for help with computer graphics and Dr. U. Reineke (JERINI BIOTOOLS GmbH) for expert advice in handling the peptide scans.

References

Cornille F, Emery P, Schüler W, Lenoir C, Mach B, Roques BP, Reith W (1998) DNA-binding properties of a chemically synthesized DNA-binding domain of hRFX1. Nucleic Acids Res 26:2143–2149

Jeltsch A, Alves J, Urbanke C, Maass G, Eckstein H, Lianshan Z, Bayer E, Pingoud A (1995) A dodecapeptide comprising the extended chain-α4 region of the restriction endonuclease *Eco*RI specifically binds to the *Eco*RI recognition site. J Biol Chem 270:5122–5129

Pingoud A, Jeltsch A (2001) Structure and function of type II restriction endonucleases. Nucleic Acids Res 29:3705–3727

Reuter M, Schneider-Mergener J, Kupper D, Meisel A, Mackeldanz P, Krüger DH, Schroeder C (1999) Regions of endonuclease *Eco*RII involved in DNA target recognition identified by membrane-bound peptide repertoires. J Biol Chem 274:5213–5221

Roberts RJ, Macelis D (2001) REBASE-restriction enzymes and methylases. Nucleic Acids Res 29:268–269

Talanian RV, McKnight J, Kim PS (1990) Sequence-specific DNA-binding by a short peptide dimer. Science 249:769–771

Talanian RV, McKnight J, Rutkowski R, Kim PS (1992) Minimum length of a sequence-specific DNA-binding peptide. Biochemistry 31:6871–6875

Wang L, Voloshin ON, Stasiak A, Stasiak A, Camerini-Otero RD (1998) Homologous DNA pairing domain peptides of RecA protein: intrinsic propensity to form β-structures and filaments. J Mol Biol 277:1–11

Suppliers

Peptide-scans

JERINI BIOTOOLS GmbH
(e-mail: biotools@jerini.de,
Internet: http://www.jerini.de,
Tel.: +49-30-63926392, Fax: +49-30-63926395)
Ost-West-Zentrum Berlin-Adlershof, Rudower Chaussee 29,
12489 Berlin, Germany

Blocking Reagent

Hoffmann-La Roche Ltd (group headquarters)
(e-mail: basel.webmaster@roche.com,
Internet: http://www.roche.de,
Tel.: +41-61-6881111, Fax: +41-61-6919391)
Grenzacherstrasse 124, Postfach, 4070 Basel, Switzerland

Affinity Purification and Competition Assays Using Solid-Phase Oligopeptides

MICHAEL MAHLER, MARTIN BLÜTHNER, and JOACHIM KOCH

Introduction

The identification and characterization of linear binding sites of monoclonal antibodies on their corresponding antigens can usually be achieved by systematic epitope-mapping studies (see Chap. 5) and mutational analyses (see Chap. 9). When analyzing polyclonal sera containing a variety of antibody specificities directed against a certain antigen, it may be necessary to distinguish between different antibody populations of the B-cell immune response. When serum antibodies recognize adjacent or overlapping epitopes, discriminating between these antibody specificities may be required. Furthermore, it is often not clear whether antibody binding to related structures is mediated by cross-reactivity of the corresponding antibodies or by different antibody populations. All these questions might be answered by using affinity-purified antibodies from adjacent or immunologically related epitopes and (or) by performing competition assays (Valle et al. 1999; Mahler et al. 2000, 2001). Also, for small-scale applications immobilized peptides can be used as an alternative

M. Mahler (✉) (e-mail: michael.mahler@pharmacia.com,
Tel.: +49-761-47805443, Fax: +49 761 47805528)
Institut für Molekulare Genetik, Universität Heidelberg
Current Address: Pharmacia Deutschland GmbH, Munzinger Straße 7,
79111 Freiburg, Germany
M. Blüthner
Institut für Molekulare Genetik, Universität Heidelberg, Germany
J. Koch
Forschungsstelle Hantaviren, Heidelberger Akademie der Wissenschaften
Current Address: Institut für Biochemie, Biozentrum N210/20,
Marie-Curie-Straße 9, 60439 Frankfurt am Main, Germany

Springer Lab Manual
J. Koch, M. Mahler (Eds.) Peptide Arrays
on Membrane Supports

to on-column purification systems for proteins and antibodies. Thus, distinct antibody specificities induced by immunization with peptides or recombinant expression fragments can easily be purified from animal sera using the target peptide bound to cellulose membranes. Since the synthesis of a huge number of different soluble peptides is time-consuming and expensive, it is obvious that solid-phase oligopeptides represent a fast, efficient and practical method for the purification of different antibody specificities.

Competition assays have proven to be a powerful tool for the elucidation of protein-protein interactions, especially for epitope-mapping studies (Blüthner et al. 1996; Wobus et al. 2000). Most of the published protocols are based on the depletion of certain antibody specificities using soluble peptides or expressed cDNA fragments as competitors that are applied in a high molar excess. The competing effects are then evaluated by a variety of immunoassays such as enzyme-linked immunosorbent assay (ELISA), Western blot, and indirect immunofluorescence (Blüthner et al. 1996; Wobus et al. 2000). Those competition assays are used to verify data derived from previous interaction studies rather than as the starting point of an interaction study. Compared to soluble competing molecules, solid-phase competitors have their individual advantages and disadvantages. When solid-phase oligopeptides are used in competition assays, epitope-specific antibodies can easily be removed from the antibody solution, whereas when soluble competitors are used, the antibody complexes either remain in the solution or have to be removed by precipitation. Without precipitation, the immunocomplexes potentially interfere with the results. Furthermore, a high number of solid-phase oligopeptides is not accessible to the antibodies due to steric hindrance. With regard to the molarity of the peptide-antigen, approximately 50 g of IgG antibodies can theoretically be removed per cm^2 of membrane-coupled target peptide. However, the actual amount is tremendously reduced by steric hindrance of the antibody molecules. Prior to each competition study, both the appropriate size of peptide-covered membrane surface and the antibody dilution have to be adjusted. In the following protocols, we describe the application of solid-phase oligopeptides for the affinity purification of epitope-specific antibodies and for competition assays.

Subprotocol 1
Affinity purification

Outline

For affinity purification of epitope-specific antibodies from polyclonal sera, oligopeptides containing the epitope sequences of interest are synthesized on activated membranes as described in Chapters 3 and 4. After blocking the nonspecific binding sites, the antibodies are allowed to react with their corresponding epitopes. Nonspecific antibodies are removed from the membrane strips by several washing steps. For the elution of specifically bound antibodies, we use the low pH method

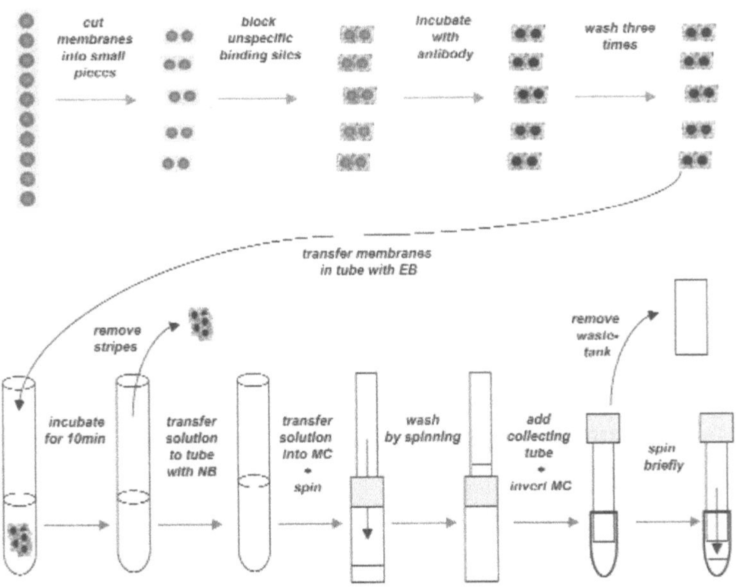

Fig. 1. Outline of the affinity purification procedure. Peptides of interest are synthesized on activated membranes (see Chap. 3 or 4). Following completion of the synthesis, membrane stripes are cut into small pieces, blocked with an appropriate blocking buffer and incubated with serum samples diluted in antibody buffer. After three washes with TBS-T, antibodies are eluted by low pH (glycine) and neutralized. Prior to use, affinity-purified antibodies are dialyzed and concentrated using Centricon-30 microconcentrators (*MC*). *EB* Elution buffer, *NB* neutralization buffer

described by Smith and Fisher (1984; also see Blüthner 1998). Alternatively, antibodies can be eluted by KSCN which works equally well (Olmstedt 1981; Krohne et al. 1982; Blüthner 1998). Following elution by low pH, the antibody solutions are neutralized by neutralization buffer. Depending on the desired application, the neutralized antibody solutions have to be dialyzed and concentrated using, e.g. Centricon-30 microconcentrators or they can be used directly. The affinity-purified antibodies can be re-incubated on nitrocellulose strips (Western blot), on peptide membranes (peptide assay), or used for indirect immunofluorescence. Furthermore, affinity-purified antibodies can be used to screen expression libraries, such as phage-display libraries (Smith 1985). The entire procedure of the affinity purification is outlined in Fig. 1.

Materials

All materials required for the synthesis of solid-phase oligopeptides are listed in Chapters 3 and 4.

Equipment

- Centrifuge (e.g. Sorvall RC 5B)
- Screw-cap tubes (15 ml)
- Centricon-30 microconcentrators (Millipore GmbH)
- Side-to-side shaker

Buffers

- Tris-buffered saline (TBS, pH 7.6):
  - 10 mM Tris/HCl
  - 150 mM NaCl
- Tris-buffered saline, 0.2 % Tween 20 (TBS-T):
  - TBS (pH 7.6)
  - 0.2 % (v/v) Tween 20
- Blocking buffer:
  - TBS (pH 7.6)
  - 0.2 % (v/v) Tween 20
  - 50 % (v/v) Inactivated horse-donor-serum
  - 150 mM sucrose
  - 10 % (v/v) 10× Blocking buffer concentrate (SU-07–250; Genosys, Cambridge, UK)
- Antibody buffer:
  - TBS

  - 0.05 % (v/v) Tween 20
  - 3 % (v/v) Inactivated horse-donor-serum
  - 150 mM sucrose
  - 10 % (v/v) 10× Blocking buffer concentrate (SU-07–250; Genosys)
- Elution buffer (pH 2.4):
  - 150 mM NaCl
  - 5 mM glycine
  - 0.1 % (v/v) Tween 20
- Neutralization buffer (pH 8.0):
  - 1 M Tris/HCl

For other blocking buffers see Chapter 11.

Procedure

The following washing steps and incubations are performed on a side-to-side shaker at room temperature.

Synthesis of peptides for the depletion reaction

1. Design and synthesize specific peptides and negative controls on activated membranes as described in Chapters 3 and 4. **Note:** The negative control peptide should have the same length and overall amino acid composition as the specific peptide, or should comprise a non-reactive sequence of the antigen.

Preparation of the membranes and antibody binding

2. Block unspecific binding sites on the membrane with blocking buffer for 2 h (see top part of Fig. 1).
3. Wash membrane strips once with TBS-T for 10 min.
4. Incubate membranes with sera diluted 1:50 in antibody buffer for 2 h. **Note:** The required volume of the serum dilution depends on the size of the membrane, on the titer of the epitope-specific antibody and on the intended application. We recommend to use at least 1 ml of serum dilution for 2 cm^2 of peptide-covered membrane surface.
5. Remove unspecific antibodies and other serum components by washing the strips three times for 5 min in TBS-T.
6. (Optional) Cut off a small piece of membrane from each peptide species and store these in TBS at 4 °C until "Monitoring of the procedure" is performed (see step 22).

Elution and neutralization of bound antibodies

7. Transfer membrane strips in a screw-cap tube filled with 1.5 ml elution buffer (see lower part of Fig. 1). **Note:** When using sizes other than 2 cm^2 of peptide surface, adjust the volumes of elution buffer and neutralization buffer equivalently.
8. Elute antibodies by incubating for 10 min, vortex tube briefly every 2 min.
9. Meanwhile, prepare a second screw-cap tube with 0.6 ml neutralization buffer.
10. Immediately neutralize the eluted antibodies by transferring the elution buffer to the second tube holding the neutralization buffer.
11. Chill antibody solutions on ice.
12. (Optional) Store membrane strips in TBS at 4 °C until "Monitoring of the procedure" is performed (see step 22).

From now on, chill antibody solutions and TBS on ice. Carry out all centrifugation steps at 4 °C.

Concentration and dialysis of eluted antibodies

13. Transfer neutralized antibody solutions in Centricon-30 microconcentrators.
14. Centrifuge the samples for 20 min at 4,000 *g*.
15. Add an appropriate volume of ice-cold TBS to the storage tank (upper part of the concentrator) (1.5–2 ml).
16. Centrifuge samples for 20 min at 4,000 *g*. **Note:** The waste tank holds a volume of about 4 ml. Therefore, the flow-through has to be removed after every two centrifugation steps.
17. Repeat steps 15 and 16 until the antibodies are washed with at least a five-fold volume of TBS compared to the initial volume loaded on the concentrator (for 1.5 ml elution buffer and 0.6 ml neutralization buffer use at least 10 ml of TBS).
18. Remove the waste tanks from the concentrators, place the collecting tube on top, and invert concentrators.
19. Spin the concentrators for 2 min at 2,000 *g* to recover the concentrated antibody solution.

20. For storage, add protease inhibitors as recommended by the manufacturers. The antibodies can be stored at 4 °C for up to 2 weeks depending on the antibody but should rather be used immediately. **Note:** Do not freeze the antibody solutions.

21. The concentrated antibodies can be directly used for indirect immunofluorescence. For Western blot and for peptide assays, dilutions of 1:20 to 1:100 in the appropriate buffer are recommended.

The following washing steps and incubations are carried out on a side-to-side shaker at room temperature.

Monitoring of the procedure (optional)

22. Incubate membrane pieces from step 6 and membrane strips from step 12 with an appropriate secondary antibody.

23. Wash membranes twice with TBS-T and twice with TBS for 5 min per wash.

24. Visualize antibody binding by using the appropriate substrate for the secondary antibody conjugate. **Note:** Peptides PRIOR to the elution should give a strong signal (see step 6), whereas peptides AFTER the elution should give no signal at all (see step 12).

Troubleshooting

Monitoring of the procedure

- No signal obtained from the membrane pieces prior to the elution (see step 6):
  Check detection system and use less stringent blocking and (or) washing conditions (see Chap. 11). If still no binding occurs, check data from previous experiments.

- Strong signal obtained from the membrane strips after the elution (see step 12):
  The elution procedure was not efficient; check the pH of the elution buffer. If the pH of the elution buffer is OK, increase elution time or stringency. Alternatively, try other elution methods.

Evaluating experiments

- The purified antibodies cause no signal in the following experiments:
  Concentration of the purified antibody is too low. If "Monitoring of the procedure" gives the desired result (see above), repeat experiments with higher antibody concentration.
- The antibody solution purified using the negative peptide causes a signal comparable to the signal of the specific purified antibodies:
  The blocking and (or) washing procedure may be not sufficient. Use a more stringent blocking buffer and (or) increase washing stringency by either increasing the number of washing steps or by using a washing buffer with a higher detergent concentration.

Subprotocol 2
Competition assays

Outline

The protein sequences of interest are synthesized as overlapping oligopeptides on activated membranes (Frank 1992; see Chaps. 3 and 4). After cutting the membranes, unspecific binding sites are blocked with blocking buffer. Afterwards, membrane stripes are washed once in TBS-T followed by incubation with the antibody (depletion reaction). The resulting antibody-depleted solutions

$$\text{Specific antibodies in percent} = \frac{(OD_{ndp} - OD_{sdp}) \times 100}{OD_{ndp}}$$	OD = optical density as measured using an ELISA-reader
	sdp = specifically depleted probe
$$\text{Residual reactivity in percent} = \frac{OD_{sdp} \times 100}{OD_{ndp}}$$	ndp = non depleted probe

Fig. 2. Calculation of specific antibody or the residual reactivity percentage

can be directly applied in the assay systems. In addition to the specifically depleted probes, as controls each competition assay should include a non-depleted sample as well as at least one sample depleted with a nonspecific peptide to ensure the specificity of the depletion reaction. When using a quantitative method for the evaluation, such as the ELISA-systems, the efficiency of the depletion reaction can either be expressed as the specific antibodies or the residual reactivity percentage. The corresponding values can be calculated indicated in Fig. 2.

The entire protocol for competition assays using immobilized peptides is outlined in Fig. 3.

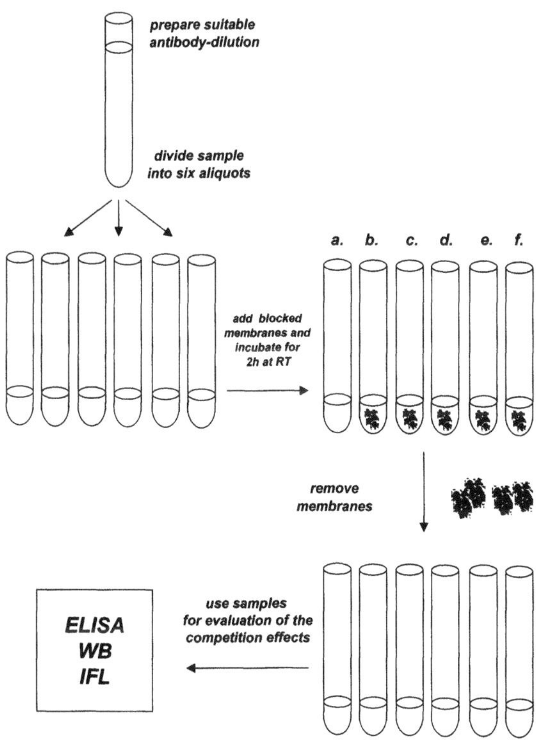

Fig. 3. Outline of a competition assay. After preparation of suitable antibody dilutions the samples is divided into (six) aliquots (depending on the number of peptides to be tested). Blocked membranes are added to the samples to be competed and incubated for 2 h at room temperature. Each competition assay should include at least one non-competed probe (*a*), one non-specific competed probe (*b*) and the specific competed probe in different amounts (*c*–*f*). After removing the membrane strips with a forceps, samples can directly be applied to immunoassays. *ELISA* Enzyme-linked immunosorbent assay, *WB* Western blot, *IFL* indirect immunofluorescence

Materials

All materials that are required for the synthesis of solid-phase oligopeptides are listed in Chapters 3 and 4. Additional materials such as blocking buffer, washing buffer and antibody buffer are listed in the affinity purification section of this chapter.

- 96-deep-well plate (optional)
- Multi-channel pipette (optional)

Procedure

Evaluation of a suitable antibody dilution

1. Prepare a dilution series of the antibody or the polyclonal serum. For sera, we recommend to start with a 1:100 dilution and to proceed with two-fold serial dilutions up to a dilution of 1:3,200 (i.e. 1:100, 1:200, 1:400...).
2. Determine the highest antibody dilution that still yields a clearly detectable signal. **Note:** The use of high antibody dilutions will reduce the required amount of the competing molecules.
3. Calculate the required volume of the antibody solution(s) for the intended experiments. **Note:** Keep in mind that, for depletion studies using ELISA system, testing of each sample in duplicate is generally recommended.

Synthesis of peptides for the depletion

4. Design and synthesize peptides under investigation and negative controls on activated membranes as described in Chapters 3 and 4. **Note:** The negative competitor should have the same length and overall amino acid composition as the specific peptide. In addition, if possible one should also use a non-reactive peptide from the antigen under investigation.

The following washing steps and incubations are carried out on a side-to-side shaker at room temperature.

Preparation of the membranes

5. Excise the peptide spots from the membranes. **Note:** Make sure that each membrane strip carries only homologous peptides. Cut each strip to the same size. Make sure that the strips are cut **PRIOR** to the blocking to avoid non-specific binding of antibodies to the cutting sites of the membrane.

6. Block nonspecific binding sites with blocking buffer for 2 h.

7. Wash membrane stripes once with TBS-T for 10 min.

Preparation of antibody dilution

8. Meanwhile, prepare the previously determined antibody dilution (see "Evaluation of a suitable antibody dilution") in antibody buffer and divide the sample into several aliquots depending on the number of different peptides to be tested (see top part of Fig. 3). **Note:** We recommend using four different peptide amounts (membrane sizes) for each peptide of interest to make sure that the competing molecules are applied in a molar excess over the sum of specific antibodies in the solution (1 cm^2 of the peptide-covered membrane should completely remove specific antibodies from 1 ml of sera diluted 1:1,000). Additionally, a non-competed probe and one probe competed with a non specific peptide should be included in the experiments. If using a large amount of different peptides, we recommend carrying out the depletion reaction in a 96-deep-well plate and to transfer probes with a multi-channel pipette to the ELISA plate.

Depletion reaction

9. Add the peptide carrying membrane stripes to the aliquots to be depleted and incubate for 2 h (see middle part of Fig. 3).

10. Remove the membrane stripes with a forceps and use depleted antibody solutions and corresponding non-depleted aliquots directly for the evaluation experiments (see lower part of Fig. 3).

11. When using a quantitative procedure, measure the competition effects and calculate the specific antibody binding and (or) the residual reactivity as indicated in Fig. 2.

Results

Biological systems

Systemic autoimmune diseases are characterized by circulating antibodies specific for defined intracellular components (Tan 1989). Sera from patients suffering from the limited form of systemic sclerosis (lSSc) have been frequently reported to contain anti-centromere antibodies (ACA; Guldner et al. 1984). In several studies it has been demonstrated that CENP-A, -B and -C represent the major targets of the anti-centromere immuno-

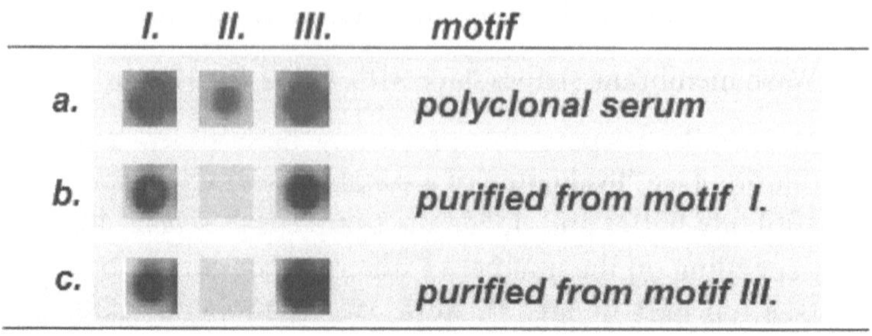

Fig. 4. Affinity purification experiment of related B-cell-epitope motifs on the centromere-protein A (Mahler et al. 2000). Affinity-purified antibodies were prepared as described in Fig. 1. Peptide-arrays carrying epitope motifs I–III of CENP-A were incubated with the polyclonal human autoimmune serum (*a*), with affinity-purified antibodies from the CENP-A-motif I (*b*) and from CENP-A-motif III (*c*)

response. Recently, we and others (Mahler et al. 2000; Muro et al. 2000) have shown that the anti-CENP-A immune response is exclusively directed against the N-terminal domain of the 17-kDa protein. Using affinity-purified antibodies we were able to demonstrate that two (I+III) of three similar sequence motifs within the N-terminal domain of CENP-A are immunologically related (Mahler et al. 2000, 2001). A simplified illustration of this experiment is presented in Fig. 4.

Another marker antigen, PM/Scl-100, is specifically recognized by a high number of sera derived from patients suffering from polymyositis-scleroderma overlap syndrome (Tan 1989; Kipnis et al. 1990; Blüthner and Bautz 1992; Ge et al. 1992). Recently, we and others were able to show that the anti-PM/Scl-100 response is predominately directed against an α-helical epitope at amino acid positions 231–245 (Ge et al. 1994, 1996; Blüthner et al. 1996; Blüthner et al. 2000). To verify these data we performed competition ELISAs using a recombinant expression fragment of the PM/Scl-100 as the target antigen and a major epitope-containing peptide covering amino acids 231–245 as its competitor. The competing reactions were carried out with a 1:1,000 serum dilution in combination with varying amounts of soluble or membrane-bound peptides (see Table 1).

When using high amounts of solid-phase (1 cm^2/ml serum dilution) or soluble (25 μg/ml serum dilution) oligopeptides of the PM/Scl-100 major-epitope sequence, most of the anti-PM/Scl-100 antibodies were depleted (see Fig. 5a). A further

Table 1. Peptide amounts for the competition reaction

Competition level	Soluble peptide (μg)	Solid-phase peptide (cm^2)
1	1	0.04
2	5	0.2
3	25	1
4	125	5

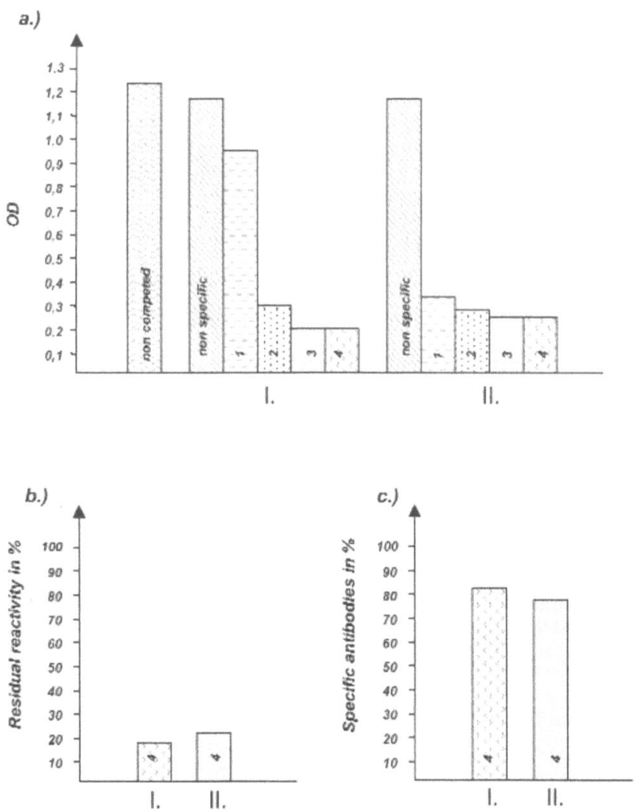

Fig. 5a–c. Result of the competition experiment using soluble (*I*) and solid-phase peptides (*II*) for the depletion reaction. The optical densities of the non-competed probe, the non-specific competed probe and the specific competed probes with the different amounts (*1–4*; see Table 1) are given in **a**. The residual reactivity (**b**) and the specific antibodies (**c**) in percent are given in Table 1

increase in molarity of the competing molecules did not increase competition effects. In contrast, both negative controls, the soluble and the membrane-bound peptides, only marginally reduced the reactivity. This indicates a specific competition effect of the PM/Scl-100 major-epitope peptide (see Fig. 5a). Thus, the findings from our previous epitope-mapping studies were confirmed. Only a minority of antibodies recognized adjacent epitopes, as indicated by the residual reactivity (17 or 23 %) of the depleted probes (see Fig. 5b) and by the specific antibodies percentage (see Fig. 5 c).

Troubleshooting

- Unspecific peptide depleted signal significantly:
  Blocking of the membranes may be insufficient. Increase blocking stringency (e.g. other blocking system; see Chap. 11).
- No competition effects of the specific probes:
  No specific binding of the antibodies. Decrease blocking and (or) washing stringency (e.g. other blocking system).
- No saturation in the competition effect with the highest amount of the competitor:
  Increase the amount of competitors and (or) use higher serum dilutions.

Acknowledgements. The experiments were supported by grant BL 483/1-1 to M.Büthner from the Deutsche Forschungsgemeinschaft. We thank Prof. Dr. E.K.F. Bautz for advice and valuable suggestions.

References

Blüthner M (1998) Methods in immunolocalization of autoantigens. In: Schenkel J (ed) RNP particles, splicing and autoimmune diseases (Springer lab manual). Springer, Berlin Heidelberg New York, pp 155–183

Blüthner M, Bautz FA (1992) Cloning and characterization of the cDNA coding for a polymyositis-scleroderma overlap syndrome-related nucleolar 100-kD protein. J Exp Med 176:973–979

Blüthner M, EKF Bautz, Bautz FA (1996) Mapping of epitopes recognized by PM/Scl autoantibodies with gene-fragment phage display libraries. J Immunol Methods 198:187–198

Blüthner M, Mahler M, Müller DB, Dünzl H, Bautz FA (2000) Identification of an alpha-helical epitope region on the PM/Scl-100 autoantigen with structural homology to a region on the heterochromatin p25beta autoantigen using immobilized overlapping synthetic peptides. J Mol Med 78: 4

Frank R (1992) Spot-synthesis: an easy technique for the positionally addressable, parallel chemical synthesis on a membrane support. Tetrahedron 48:9217–9232

Ge Q, Frank MB, Brien CO, Targoff IN (1992) Cloning of a complementary DNA coding for the 100 kD antigenic protein of the PM/Scl autoantigen. J Clin Invest 90:559–570

Ge Q, Wu Y, Trieu EP, Targoff IN (1994) Analysis of the specificity of anti-PM/Scl antibodies. Arthritis Rheum 37:1445–1452

Ge Q, Wu Y, James JA, Targoff IN (1996) Epitope analysis of the major reactive region of the 100-kd protein of PM-Scl autoantigen. Arthritis Rheum 39:1588–1595

Guldner HH, Lakomek HJ, Bautz FA (1984) Human anti-centromere sera recognise a 19.5 kD non-histone chromosomal protein from HeLa cells. Clin Exp Immunol 58:13–20.

Kipnis RJ, Craft J, Hardin JA (1990) The analysis of antinuclear and anti-nucleolar autoantibodies of scleroderma by radioimmunoprecipitation assays. Arthritis Rheum 33:1431–1437

Krohne GR, Stick R, Kleinschmidt J, Moll R, Franke WW, Hausen P (1982) Immunological localization of a major karyoskeletal protein in the nucleoli of oocytes and somatic cells of *Xenopus laevis*. J Cell Biol 94:749–754

Mahler M, Mierau R, Blüthner M (2000) Fine-specificity of the anti CENP-A B-cell autoimmune response. J Mol Med 78:460–467

Mahler M, Mierau R, Schlumberger W, Blüthner M (2001) A population of autoantibodies against a centromere-associated protein A major epitope motif cross-reacts with related cryptic epitopes on other nuclear autoantigens and on the Epstein-Barr nuclear antigen 1. J Mol Med 79:722–731

Olmsted JB (1981) Affinity purification of antibodies from diazotized paper blots of heterogeneous protein samples. J Biol Chem 256:11955–11957

Smith DE, Fisher PA (1984) Identification, developmental regulation, and response to heat shock of two antigenically related forms of a major nuclear envelope protein in *Drosophila* embryos: application of an improved method for affinity purification of antibodies using polypeptides immobilized on nitrocellulose blots. J Cell Biol 99:20–28

Smith GP (1985) Filamentous fusion phage: novel expression vectors that display cloned antigens on the virion surface. Science 228:1315–1317

Tan EM (1989) Antinuclear antibodies: diagnostic markers for autoimmune diseases and probes for cell biology. Adv Immunol 44:93–151

Valle M, Munoz M, Kremer L, Valpuesta JM, Martinez-A C, Carrascosa JL, Albar JP (1999) Selection of antibody probes to correlate protein sequence domains with their structural distribution. Protein Sci 8:883–889

Wobus CE, Hügle-Dörr B, Girod A, Petersen G, Hallek M, Kleinschmidt JA (2000) Monoclonal antibodies against the adeno-associated virus type 2

(AAV-2) capsid: epitope mapping and identification of capsid domains involved in AAV-2-cell interaction and neutralization of AAV-2 infection. J Virol 74:9281–9293

Suppliers

Amersham Pharmacia Biotech
(email: cust.orderde@eu.apbiotech.com,
Internet: http://www.apbiotech.com/na/,
Tel.: +49-800-9080713, Fax: +49-800-9080714)
Munzingerstrasse 9, 79111 Freiburg, Germany

Millipore GmbH
(Internet: http://www.millipore.com/analytical/site.nsf/home,
Tel.: +49-6196-4940, Fax: +49-6196-43901)
Hauptstrasse 87, 65760 Eschborn, Germany

Sigma-Genosys Ltd.
(e-mail: info@sigma-genosys.co.uk,
Internet: http://www.Genosys.co.uk/,
Tel.: +44-1223-839000, Fax: +44-1223-839300)
UK/Europe, London Road, Pampisford, Cambridge CB2 4EF, UK

Mutational Analysis and Structure Predictions

MARTIN BLÜTHNER, JOACHIM KOCH, and MICHAEL MAHLER

Introduction

In Chapters 3 and 4, the technique of peptide synthesis on activated membrane supports was described in detail. In this chapter, a setup will be presented that allows the researcher to obtain limited structural data regarding the peptides under investigation. The data are derived from experimental procedures combined with theoretical considerations. These procedures and theoretical considerations are described for the epitope analysis of an autoantigen. However, since antibody-antigen systems represent special cases of protein-protein interactions, the same strategies may be applied when protein-protein interactions in general are investigated.

General considerations

The most direct approach to obtain structural data about a protein of interest is the crystallization of the highly purified protein followed by X-ray diffraction or similar techniques.

M. Blüthner (✉) (e-mail: bluthner@siruius.mgen.uni-heidelberg.de,
Tel.: +49-6221-545637, Fax: +49-6221-545678),
Institute of Molecular Genetics, University of Heidelberg, INF 230,
69120 Heidelberg, Germany
J. Koch, Forschungsstelle Hantaviren, Heidelberger Akademie der Wissenschaften
Current Address: Institut für Biochemie, Biozentrum N210/20,
Marie-Curie-Straße 9, 60439 Frankfurt am Main, Germany
M. Mahler Institut für Molekulare Genetik, Universität Heidelberg
Current Address: Pharmacia Deutschland GmbH, Munzinger Straße 7,
79111 Freiburg, Germany

Springer Lab Manual
J. Koch, M. Mahler (Eds.) Peptide Arrays on Membrane Supports

There are several obstacles to this strategy. First, the protein needs to be prepared in sufficient quantities to perform such studies. Second, the protein needs to be of the highest possible purity. The purification of significant amounts of protein is often extremely laborious, and there is no simple, universal strategy to perform this task. Having purified a protein in sufficient quantities, one still has to develop strategies to crystallize the protein in such a way that it can be used for structure determination. However, one is often interested in the topology of a protein-protein interaction site, in particular in the topology of the binding between an antibody and its corresponding epitope. For such purposes it is often sufficient to develop techniques that allow the examination of the amino acids responsible for the binding in combination with theoretical structure predictions of this binding site.

Prior to the examination of a given epitope in more detail, one should consider the principal nature of an epitope. An epitope can comprise a consecutive stretch of amino acids, a structure also known as a linear or continuous epitope. Such an epitope can readily be identified with most if not all detection techniques. An epitope can also be formed by amino acids that are located at distant positions in the entire primary amino acid sequence but which come in close contact to each other once the antigen folds into its native state. Those epitopes are known as conformational or discontinuous epitopes. Most epitope-mapping techniques are based on antigens that are, at least partially, denatured and thus fail to identify conformational epitopes. However, epitopes that are clearly linear or clearly conformational represent the extremes on a scale ranging from linear to conformational structures. Local secondary structures within a given antigen are often preserved even when partially denaturing conditions are used in the assays. Such local "semi-comformational" epitopes may be detected readily under denaturing conditions. The local secondary structures may be as short as 10 amino acids. Therefore, such local secondary structures can also be adopted by individual peptides.

In this context, it is important to note that there is a fundamental difference between peptides and proteins. For the proper folding of a protein, the presence of distant amino acids on the protein may be crucial. Proteins can fold in numerous different ways, with only one defined structure representing the functional molecule. This functional structure of the protein with all its local secondary structures often is dependent on amino acids that are not located in close proximity to a given local secondary structure. Since isolated peptides lack those distant amino acids, the secondary structure of a peptide may be occasionally quite different from the corresponding structure in the native protein.

The relevance of a given amino acid within an epitope should also be considered. The contributions of the individual amino acids within an epitope can be manifold. Some amino acids may contribute directly to the binding between the epitope and its corresponding antibody. Others are responsible for the maintenance of the local secondary structure without directly interacting with the corresponding antibody (Greenspan and Di Cera 1999).

Considering the above, it becomes clear that all experimental results and theoretical musings about secondary structures should be treated with extreme care. The cumulated information should be taken as evidence rather than as proven fact. However, mutational analyses can identify the amino acids of a given epitope that are responsible for binding of the corresponding antibody. In most cases this is already the information the whole strategy was aiming at.

Outline

Experimental analysis

In principle, when the binding of an antibody to its cognate epitope is studied in detail, a first step is to identify the amino acids that are directly responsible for this interaction. This can be done by a series of mutational studies both with single amino acid substitutions as well as combinations of such mutations, as is outlined in Fig. 1. The results of the mutations first of all identify the amino acids responsible for the binding. Second, one can identify so-called mimotopes for the antibody under investigation. Mimotopes are structures that resemble a given epitope on the structural level but are different on the primary amino acid sequence level. Potentially, these results already allow for the proposal of a secondary structure.

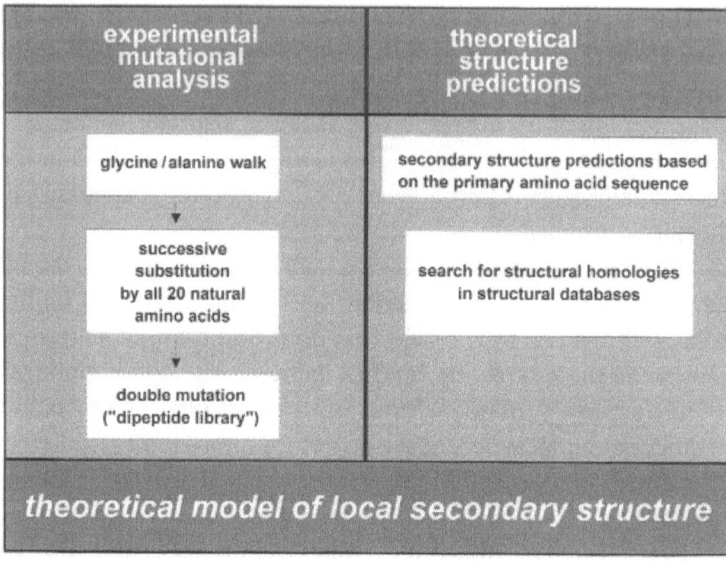

Fig. 1. Flow-chart of the experimental and theoretical strategies to obtain a theoretical model of local secondary structures as represented by the peptides under investigation

Theoretical analysis

Theoretical structure predictions can be performed in two ways. First, there are several programs that predict secondary structures based on the amino acid composition of a given peptide. Those programs mostly work with algorithms that are based on the probability of a given amino acid to be involved in either α-helices, β-strands, turns or extended structures. In addition, the adjacent amino acids are taken into account, too. Second, the amino acid sequence is compared to a structural database, where it is aligned to known protein structures. However, as yet, those structural databases contain relatively few entries, but the number of these entries is growing exponentially. The combination of both strategies should give sufficient hints for a reasonable secondary structure prediction.

Procedure

Mutational analyses: single mutations

Glycine/ alanine walk

A first step in mutational analysis is the successive replacement of the individual amino acids in the epitope by either glycine or alanine. This approach is known as glycine walk or alanine walk, respectively. The aim of this strategy is to "knock out" the amino acids that contribute to the binding between antibody and peptide. Since antibody binding is often mediated by the structural properties and the charge of amino acid side chains, glycine or alanine are used because they have no side chain (glycine) or a single methyl group as a side chain (alanine) and their net charge is neutral. As a positive control, the wild-type peptide should always be included in the setup. The number of peptides to be synthesized equals the number of amino acids contained in the peptide plus one extra peptide (the wild-type peptide):

$$N_p = N_{aa} + 1$$

where N_p is the number of peptides to be synthesized, and N_{aa} the number of amino acids of the peptide.

If there are either glycines or alanines present in the wild-type peptide, one may use a combined glycine/alanine walk, i.e. instead of a glycine at a given position in the wild-type peptide an alanine is incorporated in the mutated peptide and vice versa. Always keep in mind that, unless the peptide under investigation represents the very N-terminus of a protein, the N-terminus of the peptide should be acetylated in order to mimic its position within the context of a primary amino acid sequence. An example of such a glycine walk is given in Fig. 3.

Complete mutational analysis at single positions

In a similar approach, an amino acid at a given position is successively replaced by all of the 20 naturally occurring amino acids. In order to conveniently identify the mutated amino acids in a spatially arranged array, a given amino acid is usually replaced by all 20 amino acids, although in this way in each single case the original amino acid is synthesized again. By this approach one can identify silent mutations, i.e. mutations that preserve the antigenicity of the peptide, as well as amino acids that disrupt the binding of the antibody. The number of peptides to be synthesized thus equals the number of amino acids contained in the peptide times 20.

$$N_p = N_{aa} \times 20$$

where N_p is the number of peptides to be synthesized, and N_{aa} the number of amino acids of the peptide.

The positive control wild-type peptide is, in this approach, already contained several times due to the setup. These wild-type peptides should yield comparable intensities after visualization of bound antibodies and therefore serve as internal controls for the consistency of the peptide synthesis. An example for such a complete mutational analysis is given in Fig. 4.

Multiple mutations Having identified the amino acids that are crucial for antibody-binding, it is often a reasonable approach to introduce several mutations simultaneously. Such a setup may allow the researcher to identify mimotopes out of a pool of mutated peptides. However, this type of setup is limited by the maximum amount of peptides that can be synthesized on a membrane. The amount of peptides grows exponentially to the number of simultaneous mutations introduced.

$$N_p = 20^{N_m}$$

where N_p is the number of peptides to be synthesized, and N_m the number of simultaneous mutations.

A di-peptide library can be created routinely and requires the synthesis of only 400 peptides. By pushing the capacity of a membrane to its limits or by using high-through-put systems it may be possible to create tri-peptide libraries consisting of 8,000 peptides. However, for reasons of practicability, this represents most probably the upper limit of peptide-based libraries using the SPOT technology.

Brief procedure for mutational analysis

1. Design the peptides with the individual amino acid replacements.
2. Set up the synthesis as described in Chapters 3 and 4.

3. Detect immunoreactive peptides as described in Chapters 5 and 6.
4. Draw your conclusions.
5. Compare with the results from the theoretical considerations.

Theoretical structure predictions

Secondary structure predictions

Theoretical secondary structure predictions have been around for a quarter of a century. The early algorithms, such as those by Chou and Fasman (CF method) or Garnier, Osguthorpe and Robson (GOR method), predict secondary structures by analyzing solitary amino acid sequences (Chou and Fasman 1978; Garnier et al. 1978). More recent methods such as JPRED perform a sequence homology search and multiple alignments of the results prior to the actual prediction of secondary structures (Cuff et al. 1998). Such methods attempt to assign the query sequence to fold classes, raising the accuracy of the predictions from 50–60% as reported for the first generation methods to >70%. A good introduction to this subject can be found on the Web site of Chris Sander on the EMBL server (http://www.sander.embl-heidelberg.de/course/flow.html). An additional algorithm specifically designed for the prediction of the helical content of peptides is AGADIR (Munoz and Serrano 1994, http://www.embl-heidelberg.de/Services/serrano/agadir/ agadir-start.html).

In the case of oligopeptides even the first-generation methods should give reasonable results, since structural constraints by, in terms of the primary sequence, distant amino acids do apply for oligopeptides to a lesser degree. For practical reasons, one should always scan an amino acid sequence by several secondary structure prediction programs in order to obtain more reliable results. An excellent collection of protein analysis tools can be found on the ExPASy server (http://www.expasy.ch/tools/).

Yet, the prediction of secondary structures does not always follow exact rules. Therefore, a major aspect in the prediction of secondary structures still are the hunches of the researcher. If, for example, by any experimental method a result emerges in which relevant amino acids follow the pattern

i, $i+3$, $i+4$, $i+7$

(where i is a given residue within the primary amino acid sequence) or portions thereof, these findings are highly indicative for an α-helical region. α-Helices show a periodicity pattern of on average 3.6 amino acids per helix turn. Such an α-helical region can be deduced without the need for a secondary structure prediction program. This can be done by depicting the results by a so-called helical wheel, in which the individual residues are plotted such that the helix is viewed as an axial projection. Each amino acid residue in a helical wheel is located in a tilted position with an angle of 100° to its direct neighbor. In this way amino acid residues located at one side of the α-helix come into direct proximity in the helical wheel projection (Fig. 6).

Secondary structure prediction by search for a structural homologue

A different approach for the prediction of secondary structures is the search for a structural homologue in structure databases. We use the search option of the Automated Comparative Protein Modeling Server "SWISS-MODEL" for this purpose: http://www.expasy.ch/swissmod/SWISS-MODEL.html. The results from such a homology search also may hint at the secondary structure of the query sequence and may also serve as templates for homology modeling.

Note: The results from this type of search should be retrieved immediately. The database itself as well as the search algorithms are refined and restructured constantly. As a user of this modeling server, one does not have the option to modify these parameters. Thus, a suitable model may no longer be found on the server in later searches even when exactly the same amino acid sequence is used for the search. We have observed this phenomenon several times.

Visual display of protein and peptide structures

The coordinates of individual proteins are collected in the SWISS-MODEL database. These so-called pdb files are simple text files containing all the information necessary to display three-dimensional models of the proteins. By using an appropriate displaying program one can easily manipulate the protein models, i.e. rotate the molecule, display various features (e.g. backbone alone, or space-filling models), or assign different colors to certain secondary structures and so on. The screen images can be exported to several different graphic formats such as bitmaps, which can be easily re-imported into a graphics program for further manipulations such as residue labeling and labeling of important features.

The most widely used program for the display of structural data is RasMol (http://www.umass.edu/microbio/rasmol/). It can be downloaded free of charge from the RasMol homepage including a tutorial, a quick reference chart and an operation manual. The program is easy to use and new users can quickly become familiar with it.

Brief procedure for secondary structure predictions

1. Run secondary structure prediction programs.
2. Search for and display structural homologues in the SWISS-MODEL database.
3. Compare results with results from mutational analyses.

Results

The antigen-antibody system PM/Scl-100

The research in our group is focused on the identification of structures targeted by autoantibodies. We have investigated the interaction between the PM/Scl-100 autoantigen and its corresponding autoantibodies.

The PM/Scl-100 autoantigen is a 100-kDa protein that is part of a nucleolar multiprotein particle termed PM/Scl (Reimer et al. 1986). The PM/Scl-100 protein itself consists of 885 amino acids (accession number JH0796). Patients suffering from polymyositis-scleroderma overlap syndrome frequently develop autoantibodies against this 100-kDa protein. After cloning of the corresponding cDNA, we identified a stretch of 15 amino acids (LDVPPALADFIHQQR) at amino acid position 231–245 as the major epitope (Fig. 2; Blüthner and Bautz 1992; Blüthner et al. 1996, 2000).

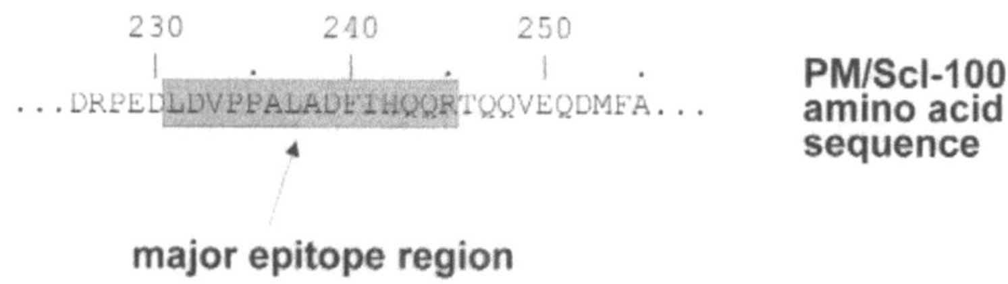

Fig. 2. The major epitope region of the PM/Scl-100 autoantigen (accession number JH0796) was identified previously between amino acids 231 and 245. The relevant amino acid sequence is *boxed* and *shaded gray* and shown within its surrounding amino acid context

As an example of mutational analyses, in this chapter the strategies are outlined that allowed us to identify those amino acids in the PM/Scl-100 major epitope that are responsible for binding of the anti-PM/Scl-100 autoantibodies. In addition, theoretical considerations are described that support the proposed secondary structure of the PM/Scl-100 major epitope.

Mutational analysis of the PM/Scl-100 major epitope

The 15-mer major epitope region of the PM/Scl-100 antigen was subjected to a mutational analysis in order to determine the amino acids relevant for the binding of the corresponding antibodies. For this purpose every single amino acid in the antigenic peptide LDVPPALADFIHQQR (amino acids 231–245) was successively replaced by glycine and the resulting mutant peptides were probed with a PM/Scl-100 positive patient serum (Fig. 3). The results show that Val-233, Pro-234, Leu-237, Phe-240, and Ile-241 of the original PM/Scl-100 sequence are the most essential amino acids for binding of the corresponding autoantibodies.

To verify this result and to identify mimotopes, a complete mutational analysis using a different anti-PM/Scl-100 serum was performed as described (Fig. 4). The central Leu-237 could only be replaced either by Ile or, to a lesser extent, by Val, two amino acids similar in size and charge. All other amino acid replacements in this position inhibited binding of the antibodies. Pro-234 could only be replaced by Val, Thr, or Ser, Phe-240 by Tyr, Trp, Met, or Leu, and Ile-241 by Val, Met, or Leu. In addition, Ala-236

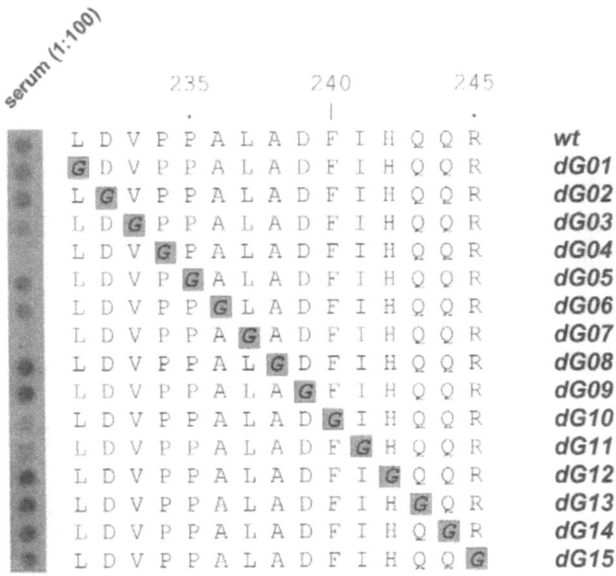

Fig. 3. Glycine walk to identify the amino acids that contribute to the binding of the antibody to the peptide. If glycine at a given position reduces or eliminates this binding, the corresponding amino acid in the wild-type peptide is directly involved in antibody binding

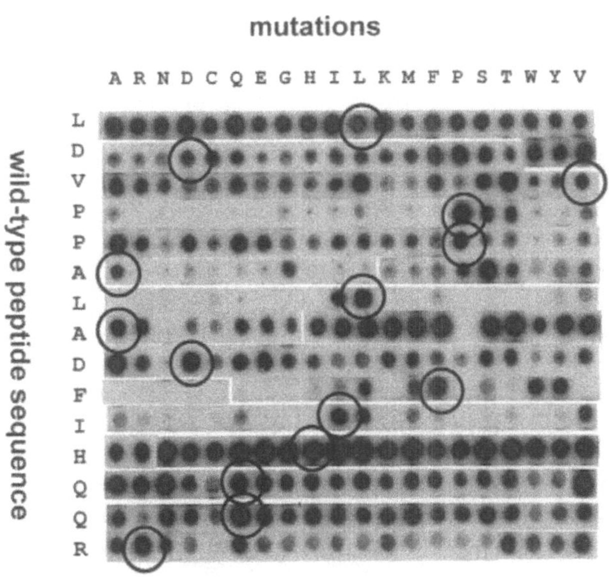

Fig. 4. Complete mutational analysis of the wild-type sequence. Each amino acid in the peptide sequence LDVPPALADFIHQQR (vertically) is replaced successively by all 20 naturally occurring amino acids (horizontally). Wild-type residues are cycled

also could be shown to contribute significantly to autoantibody binding, since it could only be replaced by Val, Tyr, Thr, Ser, Pro, or Gly. Amino acids Leu-231, Asp-232, Val-233, His-242, Gln-243, and Gln-244 obviously do not contribute to the binding of these particular antibodies, since they can be replaced by any other amino acid without effecting antigen-antibody complex formation. The results of these mutational analyses strongly suggest an α-helical conformation of these peptides.

Theoretical structure predictions for the PM/Scl-100 major epitope region

Theoretical considerations suggest a partial local α-helical secondary structure of the PM/Scl-100 major epitope according to the Chou-Fasman and Garnier-Osguthorpe-Robson methods, and by the AGADIR algorithm (Table 1, Fig. 5). It is interesting to note that, according to the Chou-Fasman method and Garnier-

Table 1. Secondary structure prediction for the PM/Scl-100 major epitope region[a]. *T* Turn, *H* α-helix

Position	Amino acid	CF-Pred[b]	GOR-Pred[c]
1	L		
2	D		
3	V		
4	P		
5	P	T	H
6	A	T	H
7	L	H	H
8	A	H	H
9	D	H	H
10	F	H	H
11	I	H	H
12	H	H	H
13	Q	H	H
14	Q	H	T
15	R	H	T

[a] As calculated by the program PEPTIDESTRUCTURE from the HUSAR program package (version 6) of the German Cancer Research Center
[b] Secondary structure prediction according to the Chou-Fasman method
[c] Secondary structure prediction according to the Garnier-Osguthorpe-Robson method

Osguthorpe-Robson methods, the α-helical region starts at amino acid position 7 (Chou-Fasman) with two preceding turns, or position 5 (Garnier-Osguthorpe-Robson), respectively, whereas AGADIR predicts the starting point for the α-helix at amino acid position 4. At positions 4 and 5 two prolines can be found. Proline is considered to be a strong helix-breaker by introducing a turn into a peptide structure. However, two such turns caused by two adjacent prolines may simulate a helix turn thus leading smoothly from an extended structure into an α-helical structure. This feature may explain the slightly different results of the various prediction algorithms. However, in principle, all the prediction algorithms agree that the PM/Scl-100 major epitope represents a local α-helical secondary structure.

A homology search with the peptide sequence LDVPPALAD FIHQQR of the PM/Scl-100 antigen against the SWISS-MODEL 3-D database revealed some topological homology with 15 amino acids (58–72) of the mouse heterochromatin protein p25β, whose 3-D structure has been reported to be determined by NMR analysis (SWISS-MODEL 3-D database, accession number 1AP0). According to these data the amino acids of the p25β protein corresponding to the antigenic peptide of the PM/Scl-100 major epitope form a local α-helix (Fig. 6).

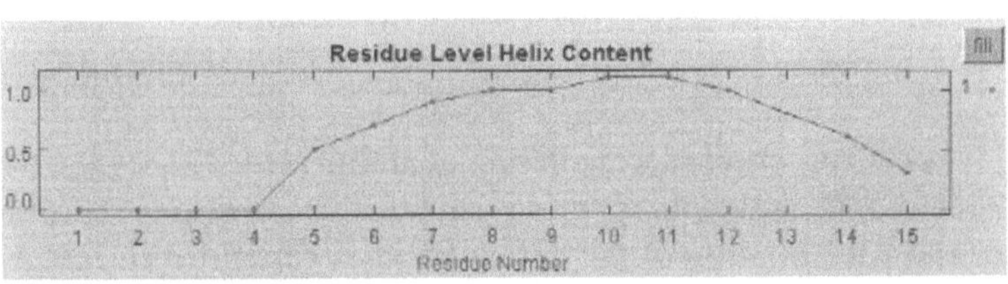

Fig. 5. Output of a secondary structure prediction by the AGADIR algorithm (see text). The algorithm predicts an α-helical structure from amino acid 4 to amino acid 15

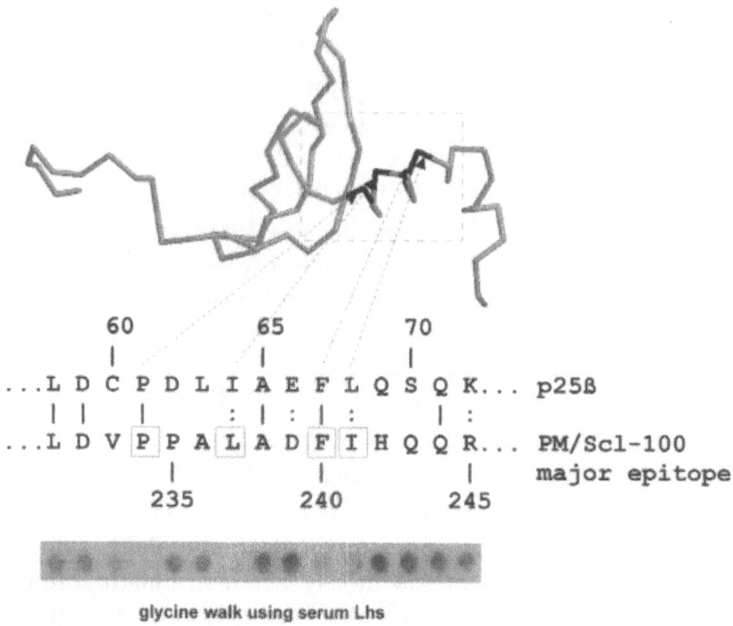

Fig. 6. Result from a homology search in the 3-D SWISS-MODEL database (see text). The query peptide LDVPPALADFIHQQR shows some structural homology to the murine heterochromatin binding protein HP1β (accession number 1AP0). Note that the amino acids of HP1βthat correspond to the immunologically relevant amino acids of the query peptide are all located on the same side of the α-helical structure of HP1β. The results from the glycine walk (see Fig. 3) are displayed below the sequences to illustrate this feature

A helical-wheel representation of amino acids 231–243 of the PM/Scl-100 antigenic peptide shows that the amino acids relevant for binding of the autoantibodies contained in human serum are in close proximity and are located on one side of the α-helix, thus rendering them accessible to an antibody (Fig. 7).

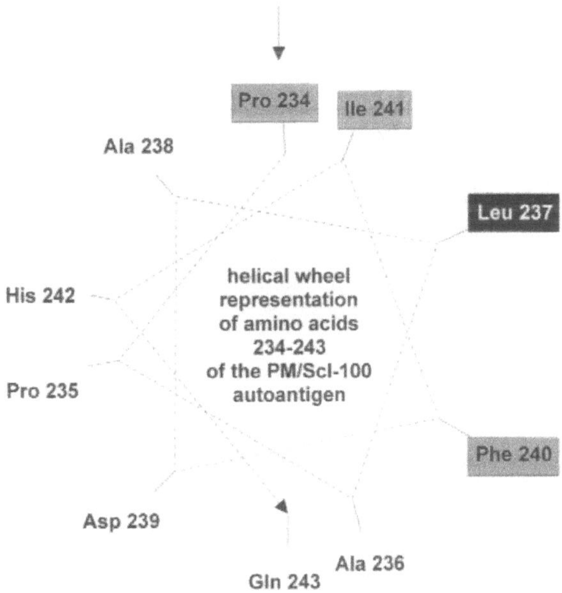

Fig. 7. Helical-wheel representation of amino acids 234–243 of the PM/Scl-100 major epitope peptide with the immunodominant amino acids highlighted. All amino acids relevant for binding of the corresponding antibody are located on the same side of the helical wheel, supporting the predicted structure

Conclusions

Using the strategies presented in this chapter, we were able to analyze the major antigenic peptide of the PM/Scl-100 autoantigen. Based on mutational analysis and theoretical structure predictions, we could propose an α-helical structure for this peptide. The results of a search for a 3-D homology suggested some structural relationship with the mouse heterochromatin protein p25. The combined results from the experimental and theoretical approaches thus show that it is possible both to determine the amino acids relevant for the binding of the peptide to its cognate antibody and to predict local secondary structures for such a peptide.

Troubleshooting

Experimental procedures

Problem	Possible Reason	Solution
Glycine/alanine walk does not identify amino acids responsible for binding (i.e. no reduction of the immunoreactivity can be observed)	The affinity of the antibody to its epitope is high enough to allow binding of the antibody even if not all relevant amino acid can contribute to the binding	Try to mutate two successive amino acids at the same time. This may reduce the affinity sufficiently to eliminate the binding between the antibody and its epitope
Contradicting results of glycine/alanine walk when using peptides of different lengths	Glycine or alanine may be too similar to the original sequence to allow reduction of antibody binding	Try an amino acid walk using a different amino acid
	Structural reasons: some amino acids may contribute not directly to the binding between antibody and epitope but rather determine or disrupt the correct folding of the peptide	Include some irrelevant spacer amino acids in the synthesis of the shorter peptides to adjust the peptide lengths

Theoretical procedures

Problem	Possible Reason	Solution
No prediction possible or not clear	Peptide too short or ambiguous	Try to use a longer peptide sequence by including adjacent amino acids
No structural homologue can be found	In the 3-D SWISS-MODEL database only a few structures can be found as compared to primary sequence databases	Use query peptides of different lengths. However, it is still possible that no structural homologue can be found

Acknowledgements. The experiments described were supported by grant BL 483/1-1 to M. Blüthner from the Deutsche Forschungsgemeinschaft.

URLs of Web pages

http://www.embl-heidelberg.de/Services/serrano/agadir/agadir-start.html

http://www.sander.embl-heidelberg.de/course/flow.html

http://www.expasy.ch/tools

http://expasy.ch/swissmod/SWISS-MODEL.html

http://www.umass.edu/microbio/rasmol

References

Blüthner M, Bautz FA (1992) Cloning and characterization of the cDNA coding for a polymyositis-scleroderma overlap syndrome-related nucleolar 100-kD protein. J Exp Med 176:973–979

Blüthner M, Bautz EKF, Bautz FA (1996) Mapping of epitopes recognized by PM/Scl autoantibodies with gene-fragment phage display libraries. J Immun Methods 198:187–198

Blüthner, M, Mahler M, Müller DB, Dünzl H, and Bautz FA (2000) Identification of an alpha-helical epitope region on the PM/Scl-100 autoantigen with structural homology to a region on the heterochromatin p25beta autoantigen using immobilized overlapping synthetic peptides. J Mol Med 78:47–54

Chou PY, Fasman GD (1978) Prediction of the secondary structure of proteins from their amino acid sequence. Adv Enzymol Relat Areas Mol Biol 47:45–148

Cuff JA, Clamp ME, Siddiqui AS, Finlay M, Barton GJ (1998) J Pred: a consensus secondary structure prediction server. Bioinformatics 14:892–893

Garnier J, Osguthorpe DJ, Robson B (1978) Analysis of the accuracy and implications of simple methods for predicting the secondary structure of globular proteins. J Mol Biol 120:97–120

Greenspan NS, Di Cera E (1999) Defining epitopes: it's not as easy as it seems. Nat Biotechnol 17:936–937

Muñoz V, Serrano L (1994) Elucidating the folding problem of helical peptides using empirical parameters. Nat Struct Biol 1:399–409

Reimer G, Scheer U, Peters JM, Tan EM (1986) Immunolocalization and partial characterization of a nucleolar autoantigen (PM/Scl) associated with polymyositis/scleroderma overlap syndrome. J Immunol 137:3802–3808

Modification of Immobilized Peptides

JOCHEN BODEM and MARTIN BLÜTHNER

Introduction

Cells need to regulate protein activity in order to control processes such as cellular metabolism and growth. This can be achieved by directly modulating protein expression or protein activity. The latter is done by posttranslational protein modifications. In addition to modifications such as acetylation, protein phosphorylation is used in vivo especially to control the activity of cellular factors. Three amino acids (serine, tyrosine and threonine) can reversibly be phosphorylated within a given context. However, not all potential phosphorylation sites are actually phosphorylated in vivo and in vitro. Several distinct pathways that transmit extracellular signals to the nucleus or to the translational machinery use phosphorylation cascades to deliver the signal to the target molecule, thereby ensuring a fast, controlled response of the cell. The S6 kinases and the Raf/MAP pathway are well-established examples of these pathways. Cell-cycle checkpoints are controlled by kinases and regulatory phosphorylations.

Furthermore, cell-cycle control pathways make use of inactivating or activating phosphorylations, e.g. to inactivate transcription factors before mitosis can occur. This allows the cells to maintain the level of certain proteins constant although their activity is not required. After mitosis these proteins are activated and there is no need for resynthesis. This demonstrates

J. Bodem (✉) (e-mail: j_bodem@med.uni-heidelberg.de,
Tel. +49-6221-561323)
Hygiene Institut, Abteilung Virologie, Im Neuenheimer Feld 324,
69120 Heidelberg, Germany
M. Blüthner
Institute of Molecular Genetics, University of Heidelberg, Germany

Springer Lab Manual
J. Koch, M. Mahler (Eds.) Peptide Arrays on Membrane Supports

the importance of understanding such regulatory pathways and events; however, to understand the cellular processes that are involved, it is necessary to dissect the modification pathways used by the cell.

Outline

The phosphorylation sites on proteins that are recognized by specific kinases have been elucidated in several reports (Tegge and Frank 1998; Dostmann et al. 1999; Himpel et al. 2000). The method presented in this chapter provides a fast and easy approach to determine the posttranslationally modified amino acid in a given protein.

We suggest the following experiments to study phosphorylation of a cloned protein of interest. As shown in Fig. 1, the first step is an in vivo labeling with ^{32}P to investigate if the protein is modified at all (Fig. 1A). For these experiments, precipitating antibodies, or at least a reasonable expression of the tagged protein in cell culture, is required. The in vivo labeling should be followed by a 2-D tryptic peptide map, in which peptides are generated by tryptic digestion of a protein followed by 2-D gel electrophoresis. The map serves as a standard for all of the in vitro experiments. These radioactive peptides also allow one to map which kind of amino acid has been modified (Boyle et al. 1991). Having obtained these results, it is necessary to establish an in vitro phosphorylation system (Fig. 1). The immobilized recombinant protein of interest is phosphorylated by cellular extracts or purified kinases followed by a subsequent tryptic phosphopeptide map to show, by co-migration with the in-vitro-labeled protein, that the same radioactive peptide pattern has been obtained (Fig. 1B). If this is the case, the modified residues must then be identified. Previously, this was a laborious and difficult task. However, the following protocol provides an outline of how oligopeptides immobilized on filters can be used to identify, or at least to narrow down, the number of modified peptides. The identified modification sites should be verified by deletion clones. In addition, identified peptides together with a computational analysis of the protein sequence allows a prediction of the possible kinases; the latter can be analyzed on immobilized peptides or protein either in vitro or in vivo by

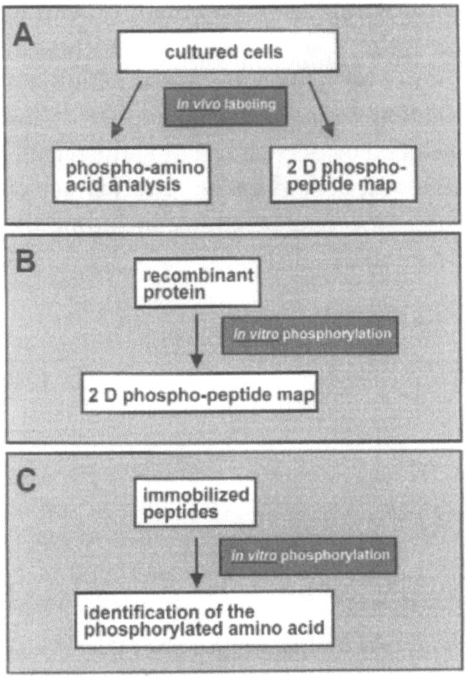

Fig. 1A–C. Analysis of protein phosphorylation. **A** In vivo phosphorylation allows the analysis of how many and which peptides are phosphorylated. **B** In vitro phosphorylation is required to proof the in vivo system. **C** Phosphorylation of immobilized peptides leads to the identification of the phosphorylated amino acid

inhibition studies. If the kinase is known, mutants which are inactive can be generated in order to discriminate the kinases involved in modulating activity from co-purifying kinase activities.

The following protocol is based on 15-mer peptides with the potential phosphorylation site at the central position. For peptides containing more than one serine, threonine or tyrosine, deletion peptides should be synthesized. We always mimic the phosphorylated and the unmodified amino acid by mutating serine into alanine (unmodified) and aspartate (addition of a negative charge mimics the phosphorylated state). This might be important if one peptide contains two phosphorylation sites that are not independently phosphorylated. Aspartate mimics the phosphorylated state and alanine ensures the phosphorylated state. After the identification of the phosphorylated residue, these mutations should be introduced in the recombinant protein to test the in vivo relevance. **Note:** If an essential amino acid is mutated, the protein might not be active after the mutation has occurred! The phosphorylation capacity of the extract or the recombinant kinase is controlled by phosphorylating the recombinant protein in a parallel reaction. If an inactive extract or kinase is available, all of the following reactions, including the controls, should be done in parallel.

Materials

Equipment

- Safety screen
- X-ray film i.e. X-Omat (Kodak)
- Tissue cultures plate 25×25
- Parafilm
- Homogenizer

Peptide scans

Peptides can either be synthesized manually (see Chap. 3) or using a pipetting robot (see Chap. 4).

Buffers and solutions

- AM100:
  - 20 mM Tris HCl, pH 8.0
  - 20 % Glycerol
  - 5 mM $MgCl_2$
  - 500 mM EDTA
  - Store at 4 °C.
- RIPA:
  - 0.5 % (w/v) SDS
  - 0.5 % (w/v) Deoxycholate
  - 100 mM NaCl
  - 20 mM Tris, pH 8
  - 10 % Glycerol
  - 10 mM EGTA
  - Store at 4 °C.

  Add protease- and phosphatase inhibitors prior to use.
- Buffer A:
  - 10 mM Tris, pH 8.0
  - 10 mM KCl
  - 1.5 mM $MgCl_2$

  Add 500 µM PMSF and 500 µM DTT before use.

- 10 × S-100 salt buffer:
  - 0.3 M Tris, pH 8.0
  - 1.4 M KCl
  - 90 mM $MgCl_2$
  - Store at 4 °C
- Buffer B:
  - 20 mM Tris, pH 8.0
  - 25 % Glycerin
  - 400 mM KCl
  - 1.5 mM $MgCl_2$
  - 1 mM EDTA, pH 8.0
  - store at 4 °C

  Add 500 µM PMSF and 500 µM DTT prior to use.
- Phosphatase inhibitors:
  - 1 M KF (store at –20 °C)
  - 20 mM Na Vanadate (store at –20 °C)
- Protease inhibitors
  - Complete (Roche, Mannheim)
- Radioactivity
  - ^{32}P-γATP (3000 mCi/ mmol)
- Ponceau S Solution (Sigma)
- Phosphorylation mix:
  - 10 µl Cellular extract (100 µg)
  - 7 µl ^{32}P-γATP
  - 4 µl KF (1 M)
  - 4 µl Na Vanadate

  Add 200 µl AM100

Cells

- 400 ml HeLa or FM3A cells (0.8–1.0 10^6 cells/ml)

Procedure

Preparation of cytoplasmic and nuclear extracts

This procedure is only required when no purified enzymes are available. When using purified enzymes, proceed with "Scanning for phosphorylation sites", step 1.

1. Spin HeLa or FM3A cells for 10 min at 200×*g* at 4 °C.
2. Wash the cell pellet with 10 ml ice-cold phosphate-buffered saline (PBS) and transfer the cell into 50-ml tubes.
3. Spin at 200×*g* for 10 min at 4 °C in a bench-top centrifuge.
4. Take off the supernatant and estimate the pellet volume (PV).
5. Resuspend the cells in buffer A (five times PV).
6. Let the cells swell for 10 min at 4 °C.
7. Centrifuge for 10 min at 200×*g* for 10 min at 4 °C.
8. Remove supernatant and resuspend the pellet in four volumes (PV) buffer A.
9. Homogenize cells with an homogenizer until the cell membranes are disrupted, but keep the nuclei intact. Monitor the success of the procedure using a microscope.
10. Centrifuge (2,000×*g* for 10 min at 4 °C) and keep the supernatant (S-100) and the pellet (nuclei).
11. Add to the supernatant 1/10 volume of 10× S-100 salt buffer.
12. Resuspend pellet in two volumes of buffer B and incubate for 30 min at 4 °C with mild agitation.
13. Centrifuge both S-100 and nuclei at 200,000×*g* in a fixed-angle rotor in an ultra-centrifuge for 45 min.
14. Dialyze supernatants against AM100 for at least 3.5 h at 4 °C until the conductivity is the same as for AM100.
15. Measure protein concentration, aliquot extracts and freeze in liquid nitrogen.
16. Store extracts at –80 °C.

Scanning for phosphorylation sites

Preparation of the membrane

1. Rinse the membrane twice with 100% ethanol. The membrane is rehydrated in AM100 for 1 h at room temperature (see Chap. 5). **Note:** Blocking is **not** required for phosphorylation.
2. Pre-wet the tissue-culture plate and place a piece of Parafilm (3 cm larger than the membrane) in the plate. Remove the remaining liquid. The Parafilm should now be fixed onto the plate. Remove the remaining liquid on the top side of the film.
3. Place the membrane on the Parafilm.

Phosphorylation

1. Add the phosphorylation mix to the membrane and incubate for 20–30 min at room temperature.
2. Remove the radioactive solution and wash the membrane at least five times with 40 ml RIPA buffer in a 50-ml Falcon tube on a side-to side shaker.
3. Control the remaining unspecific radioactivity using a hand-held monitor. If it is still too high (>300,000 cpm) continue to wash. **Note:** The signal cannot be removed by washing, since it is covalently bound.
4. Wrap the membrane in Saran wrap and estimate the exposure time counting the bound radioactivity.

Visualization

5. Expose the membrane on an X-ray film (usually <10 min)
6. Stain the membrane with Ponceau S to visualize the peptide spots. Discard the Ponceau into the radioactive waste! Align film and membrane to mark position(s) of the phosphorylated peptides. **Note:** The membrane cannot be regenerated.

Positive control reactions

1. If a purified kinase is used known target peptides should be included.
2. Recombinant protein should always be phosphorylated in parallel to control the reaction conditions. For a standard

reaction we usually use 1 μg of recombinant protein bound to 10 μl FLAG beads (Sigma, Munich, Germany). The first mock-reactions are the protein and the kinase on their own as a negative control. This will allow the differentiation of specific phosphorylations from co-purifying kinases. As a second control reaction the recombinant protein is incubated with extract or with purified kinase. The reactions volume is 20 μl and the incubation is carried out as described above. The protein beads are washed four times with 200 μl RIPA. The products are resolved by SDS-PAGE and blotted onto a nitrocellulose membrane. The membrane is then exposed on X-ray film. After aligning the film with the membrane, the radioactive band can be excised. The incorporated radioactivity should be measured in a scintillation counter and compared with previous results. This membrane can be used for a phosphopeptide map.

Results

In the following experiment, we analyzed a phosphorylation by casein kinase II (CKII). From the computer prediction we knew that seven potential sites of our protein could be phosphorylated. However, in vitro phosphorylation followed by 2-D peptide maps showed that just a single tryptic peptide was phosphorylated by the purified kinase. To identify the phosphorylated peptide, all of the potential peptides were synthesized on a membrane. The peptides were phosphorylated with recombinant kinase and cellular extracts. We could show that, although all seven peptides matched the CKII consensus sequences as predicted by the computer algorithm, only two sites were actually phosphorylated in our experiments (Fig. 2). The experiments with recombinant CKII showed exactly the same sequence specificity. With this simple approach we could eliminate five of seven possible phosphorylation sites. The two sites identified overlapped each other and localized on the same tryptic peptide. In addition, experiments with several recombinant clones allowed us finally to show that only those two sites were phosphorylated by CKII in the whole protein context. In other experiments using a similar approach but with other kinases, we could reduce the number of 35 possible phosphoserines to three that were actually

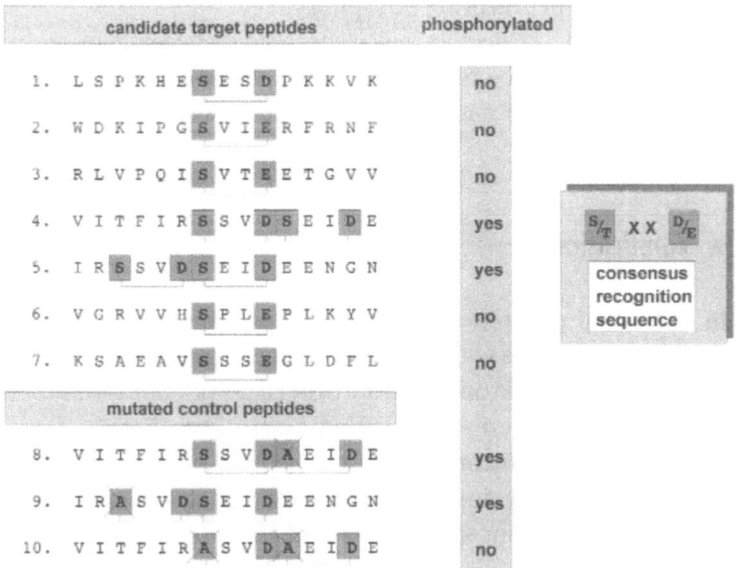

Fig. 2. Identification of in vitro phosphorylated peptides. Filters with immobilized peptides (peptides 1–7) all containing the consensus sequence [S/T]-X_2-[D/E] were phosphorylated in vitro using purified CK II. Only the overlapping peptides 4 and 5, which contain two consensus sequences, were labeled. To distinguish between the two phosphorylation sites control mutant peptides 8–10 were synthesized. As described in the text, the serine in a given peptide sequence is "knocked out" by replacement with alanine. Both peptides 8 and 9 (mutants of peptides 4 and 5) were phosphorylated, indicating that both serines represent target sites (see text). Only when both phosphorylation sites are eliminated is the peptide no longer phosphorylated

phosphorylated by the recombinant kinase, underlining the power of this method especially when the consensus sequences of a phosphorylation site is unknown.

Troubleshooting

- Control reaction
  a) Incorporation into the control reaction is low.
  Add "cold" ATP to the reaction to reach the optimal K_m
  b) Negative control is phosphorylated.
  If possible, use magnetic beads to bind the target protein, since these are less "sticky."

- Too many peptides are phosphorylated.
  If the your protein is well-conserved between mouse and human and if the regulation is not species-dependent it may be worth eliminating unconserved phospho-amino acid residues.

Applications

In additional experiments, it was also possible to acetylate the side chains of lysine within a peptide sequence. However, there is still a requirement to optimize the experimental conditions for in vitro acetylations. In another approach, pre-phosphorylated amino acids were incorporated in a peptide synthesis. A rabbit antiserum raised against this peptide was capable of discriminating between the phosphorylated and the non-phosphorylated peptide.

References

Boyle WJ, van der Geer P, Hunter T (1991) Phosphopeptide mapping and phosphoamino acid analysis by two-dimensional separation on thin-layer cellulose plates. Methods Enzymol 201:110–49

Dostmann WR, Nickl C, Thiel S, Tsigelny I, Frank R, Tegge WJ (1999) Delineation of selective cyclic GMP-dependent protein kinase Ialpha substrate and inhibitor peptides based on combinatorial peptide libraries on paper. Pharmacol Ther 82:373–387

Himpel S, Tegge W, Frank R, Leder S, Joost HG, Becker W (2000) Specificity determinants of substrate recognition by the protein kinase DYRK1A. J Biol Chem 275:2431–2438

Tegge WJ, Frank R (1998) Analysis of protein kinase substrate specificity by the use of peptide libraries on cellulose paper (SPOT-method). Methods Mol Biol 87:99–106

Suppliers

Hoffmann-La Roche Ltd (Group Headquarters)
(e-mail: basel.webmaster@roche.com,
Internet: http://www.roche.com,
Tel.: +41-61-688 1111, Fax: +41-61-691 9391)
Grenzacherstrasse 124, Postfach, 4070 Basel, Switzerland

Sigma-Aldrich Chemie GmbH
(e-mail: deorders@eurnotes.sial.com,
Internet: http://www.Sigma-aldrich.com/,
Tel.: +49-89-65130, Fax: +49-89-65131169)
Eschenstrasse 5, 82024 Taufkirchen, Germany

Immobilized Peptides to Study Protein-Protein Interactions – Potential and Pitfalls

RÜDIGER BRÄUNING, MICHAEL MAHLER,
BARBARA HÜGLE-DÖRR, MARTIN BLÜTHNER,
JOACHIM KOCH, and GABRIELE PETERSEN

Introduction

In recent years, immobilized peptides synthesized on activated cellulose membranes (SPOT synthesis, Frank 1992) have become an important tool in the study of protein-protein interactions and numerous other aspects of molecular recognition (reviewed in Reineke et al. 2001). A broad range of applications has already been described and the rising interest in proteomics is bound to rely on the enormous potential of the (automated) method. When adapted to high-throughput screening, SPOTs, i.e. peptide arrays, will become an invaluable tool in pharmacogenomics and drug discovery.

This chapter deals with the technicalities of probing membrane-bound peptides with protein targets on a more moderate scale, focusing on monoclonal antibodies and polyclonal human sera, the effects of protein concentration, blocking conditions, detection systems, and the efficiency of the regeneration procedure.

M. Mahler, M. Blüthner, J. Koch, G. Petersen (✉)
(e-mail: gabriele.petersen@urz.uni-heidelberg.de,
Tel.: +49-6221-545256, Fax +49-6221-545678)
Institute of Molecular Genetics, University of Heidelberg,
Im Neuenheimer Feld 230, 69120 Heidelberg, Germany
Current address: J. Koch, M. Mahler see Chapters 6 and 8
Current address: R. Bräuning, German Cancer Research Center (DKFZ),
Im Neuenheimer Feld 280, 69120 Heidelberg, Germany
Current address: B. Hügle-Dörr, SCICUS, Zeppelinstrasse 12,
69168 Wiesloch, Germany

Springer Lab Manual
J. Koch, M. Mahler (Eds.) Peptide Arrays on Membrane Supports

General considerations

Blocking conditions

As known for other protein-binding assays on a solid surface, prior to incubation with a putative interaction partner, unspecific binding-sites on membranes harboring synthesized peptides have to be blocked by incubation with blocking solutions.

In general, blocking agents can be divided into several groups. The first group uses non-ionic detergents such as Tween-20 or Triton X-100 to reduce charge-specific interactions. A second group consists of proteins such as bovine serum albumin (BSA), milk casein (usually in the form of skim milk powder) or sera of different species (Vogt et al. 1987; Peterfi and Kocsis 2000). A third group of blocking solutions combines non-ionic and protein-based systems and sometimes gelatin. Recommendations given in published protocols are often contradictory and are highly dependent on the object of investigation. For Western immunoblotting, Tween-20 has been shown to reduce non-specific binding and background signals when used at 4 °C (Thean and Toh 1989). However, in other experiments Tween-20 was found to lead to high non-specific binding (Kenna et al. 1985) or decreased the binding of specific autoantibodies (Zampieri et al. 2000). Some authors warn against using Tween-20 because it can remove proteins from nitrocellulose membranes (Hoffmann and Jump 1986).

Regarding ELISAs, bovine serum albumin and casein showed greater blocking efficiency than other frequently recommended blocking agents (Kenna et al. 1985; Pruslin 1991), whereas protein-based agents were also found to lower the reactivity between the antibody and the antigen when compared to Tween-20 (Mohammad and Esen 1989).

Interacting proteins – antibody source

To determine the epitope recognized by an antibody using membrane-bound peptide arrays, primary antibodies can be:

- Purified monoclonal antibodies
- Hybridoma cell culture supernatant
- Polyclonal sera

Other sources, e.g. recombinant antibodies in the form of Fabs, single-chain antibody fragments (scFv), or filamentous phage displaying scFv, may be more difficult to analyze. Even if a scFv specifically recognizes its original protein target, scFv have been found to display a promiscuous reaction with completely unrelated peptides. Although promiscuity can also be observed with monoclonal antibodies and polyclonal sera (Kramer et al. 1997; Wobus et al. 2000), this behavior is much more pronounced with scFv and makes the unambiguous determination of the "real" epitope sometimes almost impossible. Phage particles show tremendous unspecific affinity to many peptide sequences and, to a lesser extent, to cellulose membranes, resulting in high non-specific binding that can hardly be reduced, regardless of the blocking system.

Detection assays

In epitope mapping, the primary antibody recognizing its membrane-bound peptide is generally detected with a secondary antibody directed against the constant domains of the primary antibody. The secondary antibody is conjugated to an enzyme moiety, usually horseradish peroxidase or alkaline phosphatase. The resulting complex is then visualized by the addition of a substrate, which is converted by the enzyme into a reaction product generating a visible signal in the form of a colored precipitate or by the emission of light (chemiluminescence). Alternatively, enzyme-linked protein A or protein G complexes may be used.

If the interacting protein is not an antibody, there are several other options:

- An antibody specifically recognizing the interacting protein is available.
- The protein is produced recombinantly and contains an engineered "tag", which in turn is recognized by a specific antibody.
- If no secondary detection system is available, the protein itself must be labeled. This is most easily achieved either by incorporation of radioactive amino acids during synthesis or by biotinylation of free amino groups.

In the first two cases, detection follows the basic protocol for antibodies. In the latter case, radioactively labeled proteins will be visualized directly by autoradiography. Biotinylated interaction partners are detected following incubation with avidin or streptavidin usually conjugated to horseradish peroxidase. Avidin and streptavidin have an extraordinary affinity for biotin, which is almost irreversible. Moreover, depending on the number of biotins present in the protein, the system can be used to enhance weak signals. However, it must be determined that the addition of biotins does not interfere with the protein-protein interaction to be detected.

If a membrane (see below) is to be reused, systems based on colored precipitates and biotinylated molecules should be avoided, since these cannot be removed from the membrane.

Membranes can be regenerated and subjected to further use with a second antibody or interaction partner. To generate unambiguous results, the interaction partner from the first probing must be completely removed from the peptide spot or the membrane, respectively.

Procedure

Depending on the peptide interaction partner (e.g. antibodies, proteins or DNA), different regeneration procedures may be applied (see also Chaps. 5, 6 and 7). For antibodies the following protocol is recommended.

Following three short washes in, respectively, water, dimethylformamide, and water, membranes are incubated in regeneration buffer A [8 M urea, 1 % (w/v) SDS, 0.1 % (v/v) β-mercaptoethanol] for 3×10 min and in regeneration buffer B [50 % (v/v) ethanol, 10 % (v/v) acetic acid] for 3×10 min, rinsed in ethanol and dried. This commonly used regeneration procedure for antibodies is based on three elution principles:

- Denaturing conditions (buffer A)
- High salt (buffer A)
- Low pH (buffer B)

Three washing steps (10 min each) under denaturing conditions and high salt followed by three washes at low pH (2–3) should be sufficient to disrupt even the strongest protein-protein interaction; however, removal of the interaction partner from the membrane is not 100 % and depends very much on the blocking system and the amount of protein used in the first round of incubation.

Materials

Immobilized peptides

Peptide synthesis is described in Chapter 3 (manual synthesis) and Chapter 4 (automated synthesis).

Blocking solutions

For a comparative study (see below) to assess the efficacy of different blocking buffers on cellulose membranes, we have selected four frequently used blocking solutions for immunoassays and one SPOT-specific system all prepared in TBS (10 mM Tris/HCl, pH 7.6, 150 mM NaCl). For interaction studies with antibodies TBS works well; however, if the reaction partner calls for a different buffer, other commonly used buffer systems are compatible with cellulose membranes.

- Bovine serum albumin (BSA) block (I)
  5 % (w/v) BSA
  0.2 % (v/v)Tween-20 in TBS
  For antibody incubation, BSA was reduced to 1 % and Tween-20 to 0.05 %.
- Horse serum block (II)
  50 % (v/v) Horse serum
  0.2 % (v/v) Tween-20 in TBS
  For antibody incubation, Tween-20 was reduced to 0.05 %.
- Milk block (III)
  5 % (w/v) Skim milk powder
  0.2 % (v/v) Tween-20 in TBS
  For antibody incubation, skim milk powder was reduced to 1 % and Tween-20 to 0.05 %.
- Tween-20 block (IV)
  0.2 % (v/v) Tween-20 in TBS
  For antibody incubation, Tween-20 was reduced to 0.05 %.
- Superblock (V)
  50 % (v/v) Horse serum
  10 % (v/v) 10× Blocking buffer concentrate (SU-07–250, 10×; Genosys, Cambridge, UK)
  0.2 % (v/v) Tween-20
  150 mM Sucrose in TBS
  For antibody incubation, Tween-20 was reduced to 0.05 % and horse serum to 3 %.

Procedure

Epitope mapping and blocking systems

We have assessed the efficacy of the five above-mentioned blocking systems in a comparative study with a monoclonal antibody and a polyclonal serum. The monoclonal antibody BP53-11 is directed against the tumor suppressor protein p53 and recognizes the N-terminal epitope FSDLWKL (Table 1; Fack et al. 1997). Comparative peptide immunoassays were performed with four antibody concentrations (1 μg/ml, 0.2 μg/ml, 40 ng/ml, 8 ng/ml) diluted in the respective assay buffer. As polyclonal serum, a human autoimmune serum derived from a patient suffering from the limited form of systemic sclerosis was selected. This serum contains high titers of anti-centromere antibodies including antibodies specific for CENP-A. One well-defined epitope (KPEAPRRR, Table 1) is located within the first antigenic region of the N-terminal domain of centromere protein A (Mahler et al. 2000). The serum was used in dilutions of 1:100, 1:200, 1:400, and 1:800.

Secondary antibodies were peroxidase conjugated goat-anti-human IgG or goat-anti-mouse IgG+IgM (Dianova, Hamburg, Germany).

Peptide immunoassays were performed according to previously published protocols (Fack et al. 1997, Blüthner et al. 2000; Mahler et al. 2000, Chap. 5 of this manual). In brief: following overnight blocking at 4 °C in the different blocking buffers, membranes were washed in TBS/0.05 % (v/v) Tween-20 and incubated for 2 h at room temperature with the different antibodies diluted in the respective assay buffers. Unbound antibodies were removed by three washes (5 min each) in

Table 1. Amino acid sequences of the synthesized peptides with respect to the positions in the protein recognized by the two antibodies. Core epitopes are indicated in bold

Antibody	Epitope-bearing peptide	Control
BP53-11	QET**FSDLWKL** (p53 amino acids 19–25)	LSQETFSDLW (p53 amino acids 17–23)
CEN-0010	**KPEAPRRR**SP (CENP-A amino acids 9–16)	PRRRRSPSP (CENP-A amino acids 13–20)

TBS/0.05% (v/v) Tween. Following a 1-h incubation with peroxidase-conjugated secondary antibodies (dilution 1:5,000 in the respective assay buffers), reactive spots were visualized by enhanced chemiluminescence (ECL, Amersham Pharmacia Biotech, Freiburg, Germany)

Following visualization, the membranes used in our comparative study were regenerated according to the procedure described above (see General considerations, Regeneration). After regeneration they were blocked in the very same blocking buffers using the same incubation conditions as those for the immunoreaction.

Results

Results are summarized in Fig. 1A. Monoclonal antibody BP53-11 recognizes the spot containing its epitope sequence, whereas no reaction was detected with a control peptide synthesized according to amino acids adjacent to the epitope. Incubation of the secondary goat-anti-mouse antibody alone indicated that

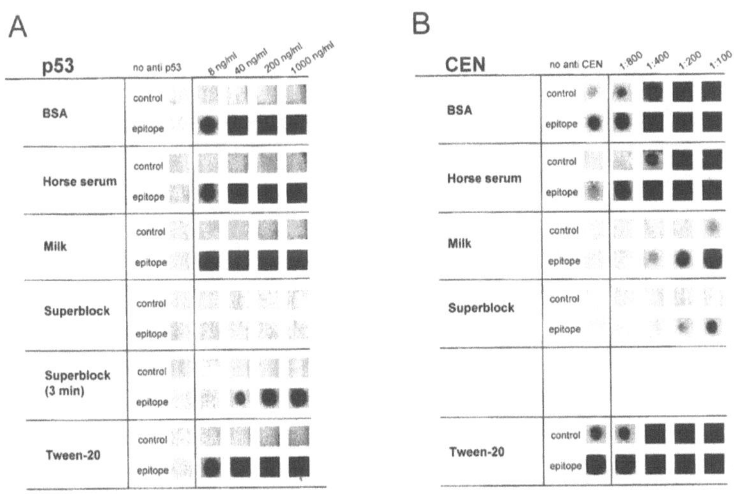

Fig. 1. Membranes harboring epitope and control peptides (see Table 1) were incubated with four different antibody dilutions (indicated at the *top*). As a control, one set of peptides was incubated without the primary antibody. Films were exposed for 3 s for anti-p53 (**A**) or 10 s for CEN (**B**). A 3-min exposure time for super-blocked membranes is also shown

this antibody preparation does not interact with these peptides in the absence of the primary antibody. Following a 3-s exposure, all antibody concentrations yielded strong signals on the epitope spots of membranes blocked with buffers I–IV, i.e. bovine serum albumin, horse serum, milk, and Tween-20, whereas membranes blocked with Superblock had to be exposed for at least 3 min. Even then, signal intensities were much lower, no signal could be detected with an antibody concentration of 8 ng/ml and the highest antibody concentration gave results comparable to the lowest antibody concentrations using the other blocking buffers. Prolonged exposure also showed that, with the exception of Superblock, higher antibody concentrations resulted in increased background staining of the membrane.

If an unknown epitope sequence is mapped, the protein sequence of interest is dissected into overlapping peptides. Thus, dependent on the length of the epitope and the peptides synthesized, several adjacent spots will react with the respective antibodies. When blocking buffers I–IV, were used, it became difficult to discern individual adjacent reactive spots on an autoradiography after a 3-s exposure if the antibody concentration rose above 40 ng/ml. For membranes blocked with Tween-20 and, most prominently, with horse serum, longer exposures yielded no discernible spots, and everything turned black independent of the antibody concentrations employed.

As shown in Fig. 1B, the secondary horseradish-peroxidase-coupled anti-human antibody reacted strongly with the control and specific peptides in the absence of the primary serum. This unspecific reaction was completely eliminated when reactions were carried out in Superblock. Blocking with milk was sufficiently effective when higher serum dilutions were applied, whereas the other blocking systems were inefficient.

Although it is necessary to adjust the blocking and detection system to the specific application, several general considerations emerged from these results:

- Always check the reactivity of the secondary detection system; adjust the blocking system and the dilution of the secondary antibody accordingly.
- For strong monoclonal antibodies, the blocking system is not critical if the secondary detection does not result in unspecific background staining.

- Due to the high local concentration of epitope sequences present in a single spot, the antibody concentration can be kept very low. This will: (1) save material if the antibody preparation is limiting and large incubation volumes, as for a complete membrane, are needed, and (2) it will ensure that individual spots remain discernible.
- Effective blocking reduces all background signals and will result in a clear autoradiogram when using ECL; it may thus be difficult to assign the reactive spots to the respective peptides. Upon prolonged exposures, the outline of the membrane and some background staining may appear. However, to unambiguously correlate reactive spots and epitope sequence, mark the outline of the membrane on the film, prepare a grid according to the spot's grid of the membrane on a transparent film, superimpose, align, and identify the reactive spots.
- Polyclonal sera contain a large number of antibody specificities, which require strong blocking systems. As a consequence, it might not be possible to detect specific reactivities due to low affinities or low titers of the specific antibody.
- Strong blocking system can also block specific interaction sites, which may be counteracted by increasing the amount of interaction partners in the assay or by prolonged exposure times when using chemiluminescence-based detection systems.
- In our hands, Superblock presented itself as the most effective blocking system. Milk is not quite as strong but probably good enough for most assays, whereas horse serum, BSA, and especially Tween-20 were not very effective in reducing unspecific reactions in peptide immunoassays. As a consequence, they also do not interfere strongly with the specific interaction. Thus, if the interaction partner including the detection system is devoid of unspecific interactions, moderate blocking is sufficient, especially when low affinities are expected and small amounts of interaction partners are present.

We regenerated the membranes used in our comparative study (see above) and blocked them in the very same blocking buffers.

The subsequent incubation was performed with the detection system alone, omitting the primary antibodies. Although signal intensities were strongly reduced, reactive spots were still visible even after short exposure times. Prolonged exposure clearly showed that regeneration efficiency is directly correlated with blocking efficiency. Membranes blocked with Superblock were completely devoid of any signal. Second best turned out to be horse serum, closely followed by milk where a 15-min exposure was necessary to make a negligible signal visible on membranes that had been incubated with the highest antibody concentration. This blocking capacity of horse serum, however, did not compensate for the disadvantage of unspecific reactions with the membranes. Milk was not as good as Superblock, but remaining protein could only be detected on membranes incubated with the highest antibody concentration. In contrast, Tween-20 and BSA blocks yielded quite strong signals at antibody concentrations of 1 µg/ml and 200 ng/ml; signals from 40 ng/ml were as negligible as for horse serum at 1 µg/ml.

References

Blüthner M, Mahler M, Müller DB, Dünzl H, Bautz FA (2000) Identification of an alpha-helical epitope region on the PM/Scl-100 autoantigen with structural homology to a region on the heterochromatin p25beta autoantigen using immobilized overlapping synthetic peptides. J Mol Med 78:47–54

Fack F, Hügle-Dörr B, Song D, Queitsch I, Petersen G, Bautz EKF (1997) Epitope mapping by phage display: random versus gene-fragment libraries. J Immunol Methods 206:43–52

Frank R (1992) SPOT synthesis: an easy technique for the positionally addressable, parallel chemical synthesis on a membrane support. Tetrahedron 48:9217–9232

Hoffmann WL, Jump AA (1986) Tween-20 removes antibodies and other proteins from nitrocellulose. J Immunol Methods 94:191–196

Kenna JG, Major GN, Williams RS (1985) Methods for reducing non-specific antibody binding in enzyme-linked immunosorbent assays. J Immunol Methods 85:409–419

Kramer A, Keitel T, Winkler K, Stöcklein W, Höhne W, Schneider-Mergener J (1997) Molecular basis for the binding promiscuity of an anti-p24 (HIV-1) monoclonal antibody. Cell 91:799–809

Mahler M, Mierau R, Blüthner M (2000) Fine-specificity of the anti CENP-A B-cell autoimmune response. J Mol Med 78:460–467

Mohammad K, Esen A (1989) A blocking agent and a blocking step are not needed in ELISA, immunostaining dot-blots and western blots. J Immunol Methods 117:141–145

Peterfi Z, Kocsis B (2000) Comparison of blocking-agents for an ELISA for LPS. J Immunoassays 21:341–354

Pruslin FH (1991) Caveats and suggestions for the ELISA. J Immunol Methods 137:27–35

Reineke U, Volkmer-Engert R, Schneider-Mergener J (2001) Applications of peptide arrays prepared by the SPOT-technology. Curr Opin Biotechnol 12:59–64

Thean ET, Toh BH (1989) Western immunoblotting: temperature-dependent reduction in background staining. Anal Biochem 177:256–258

Vogt RF Jr, Phillips DL, Henderson LO, Whitfield W, Spierto FW (1987) Quantitative differences among various proteins as blocking agents for ELISA microtiter plates. J Immunol Methods 101:43–50

Wobus C, Hügle-Dörr B, Girod A, Petersen G, Hallek M, Kleinschmidt JA (2000) Monoclonal antibodies against the adeno-associated virus type 2 (AAV-2) capsid: epitope mapping and identification of capsid domains involved in AAV-2-cell interaction and neutralization of AAV-2 infection. J Virol 74:9281–9293

Zampieri S, Ghirardello A, Doria A, Tonello M, Bendo R, Rossini K, Gambari PF (2000) The use of Tween-20 in immunoblotting assays for the detection of autoantibodies in connective tissue diseases. J Immunol Methods 239:1–11

Suppliers

ABIMED Analysen-Technik GmbH
(Tel: +49-21-73-8905-0, Fax: +49-21-73-890577)
Raiffeisenstrasse 3, 40764 Langenfeld, Germany

Amersham Pharmacia Biotech
(e-mail: cust.orderde@eu.apbiotech.com,
Internet: http://www.apbiotech.com/na/,
Tel.: +49-800-9080713, Fax: +49-800-9080714)
Munzingerstrasse 9, 79111 Freiburg, Germany

Dianova GmbH (e-mail: info@dianova.de,
Tel.: +49-40-450670, Fax: +49-40-45067490)
Mittelweg 176, 20148 Hamburg, Germany

Sigma-Genosys Ltd. (e-mail: info@sigma-genosys.co.uk,
Internet: http://www.Genosys.co.uk/,
Tel.: +44-1223-839000, Fax: +44-1223-839300)
UK/Europe, London Road, Pampisford, Cambridge CB2 4EF, UK

Abbreviations

Ab	antibody
Acm	amidomethyl group
ACA	anti-centromere antibodies
ATP	Adenosin 5'-Triphosphate
BOC	tert-butyloxycarbonyl group
BPB	bromophenol blue
BSA	bovine serum albumine
CDR	complemetarity determining region
CENP-A	centromer protein A
CF method	Chou and Fasman method
CK II	casein kinase II
DCM	dichloromethan
DIC	diisopropylcarbodiimide
DKP	diketopiperazine
DMF	dimethyl formamide
DNA	deoxyribonucleic acid
DTT	1,4-dithio-DL-threitol
EB	elution buffer
ECL	enhanced chemiluminescence
ELISA	enzyme-linked immunosorbent assay
Fab	antibody fragment ab
Fmoc	9-fluorenylmethyloxycarbonyl group
GOR method	Garnier, Osguthrope and Robson method
GTP	Guanosin 5'-Triphosphate
HBr	hydrogen bromide
HF	hydrogen fluoride
HOBt	1-hydroxybenzotriazole
HPLC	high performance liquid chromatography
HRP	horseradish peroxidase
IFL	indirect immunofluorescence
IgA	immunnoglobulin A

IgD	immunnoglobulin D
IgG	immunnoglobulin G
IgE	immunnoglobulin E
IgM	immunnoglobulin M
kDa	kilo Dalton
KSCN	potassium thiocyanate
LHC	light-harvesting complex
lSSc	limited form of systemic sclerosis
mAb	monoclonal antibody
MC	microconcentrator
MHC	major histocompatibilty complex
MS	mass spectrometry
NB	neutralization buffer
NMP	1-methyl-2-pyrrolidone
OD	optical density
OLC	one-letter code
Opfp	pentafluorophenyl ester
ORF	open reading frame
PAGE	polyacrylamide gelelectrophoresis
Pbf	2,2,4,6,7-pentamethyldihydrobenzofuran-5-sulfonyl group
PEG	polyethylene glycol
PM/Scl-100	polymyositis/sclerodera overlapping syndrome associated antigen 100 kDa
PV	pellet volume
rpm	revolutions per minute
scFv	single chain Fv antibody
SDS	sodium dodecylsulfate
SPPS	solid-phase peptide synthesis
SRP	signal recognition particle
StBu	tert-butyhlthio group
TBS	tris-buffered saline
TBS-T	TBS-Tween 20
TFA	trifluoroacetic acid
THF	tetrahydrofuran
Tris	Tris-hydroxymethyl-aminomethan
Trt	trityl group
v/v	volume/(total)-volume
w/v	weight/(total)-volume
WB	western blot

Subject Index

FSC
www.fsc.org
MIX
Papier aus verantwortungsvollen Quellen
Paper from responsible sources
FSC® C105338

If you have any concerns about our products,
you can contact us on
ProductSafety@springernature.com

In case Publisher is established outside the EU,
the EU authorized representative is:
Springer Nature Customer Service Center GmbH
Europaplatz 3, 69115 Heidelberg, Germany

Printed by Libri Plureos GmbH
in Hamburg, Germany